Functional Verification Coverage Measurement and Analysis

Andrew Piziali

Functional Verification Coverage Measurement and Analysis

Andrew Piziali

Library of Congress Control Number: 2007932007

ISBN 978-0-387-73992-2 e-ISBN 978-1-4020-8026-5

Printed on acid-free paper.

9 8 7 6 5 4 3 2 1

springer.com

Table of Contents

Foreword ix

Preface xiii

Introduction 1

1. The Language of Coverage 5

2. Functional Verification 15
- 2.1. Design Intent Diagram 16
- 2.2. Functional Verification 17
- 2.3. Testing versus Verification 19
- 2.4. Functional Verification Process 19
  - 2.4.1. Functional Verification Plan 20
  - 2.4.2. Verification Environment Implementation 26
  - 2.4.3. Device Bring-up 27
  - 2.4.4. Device Regression 28
- 2.5. Summary 30

3. Measuring Verification Coverage 31
- 3.1. Coverage Metrics 31
  - 3.1.1. Implicit Metrics 32
  - 3.1.2. Explicit Metrics 33
  - 3.1.3. Specification Metrics 33
  - 3.1.4. Implementation Metrics 34
- 3.2. Coverage Spaces 34
  - 3.2.1. Implicit Implementation Coverage Space 35
  - 3.2.2. Implicit Specification Coverage Space 35
  - 3.2.3. Explicit Implementation Coverage Space 36
  - 3.2.4. Explicit Specification Coverage Space 37
- 3.3. Summary 38

4. Functional Coverage 39
- 4.1. Coverage Modeling 39
- 4.2. Coverage Model Example 40
- 4.3. Top-Level Design 44

4.3.1. Attribute Identification 45
4.3.2. Attribute Relationships 50
4.4. Detailed Design 61
4.4.1. What to Sample 62
4.4.2. Where to Sample 65
4.4.3. When to Sample and Correlate Attributes 66
4.5. Model Implementation 67
4.6. Related Functional Coverage 75
4.6.1. Finite State Machine Coverage 75
4.6.2. Temporal Coverage 76
4.6.3. Static Verification Coverage 77
4.7. Summary 78

5. Code Coverage 79
5.1. Instance and Module Coverage 79
5.2. Code Coverage Metrics 80
5.2.1. Line Coverage 80
5.2.2. Statement Coverage 81
5.2.3. Branch Coverage 82
5.2.4. Condition Coverage 84
5.2.5. Event Coverage 84
5.2.6. Toggle Coverage 85
5.2.7. Finite State Machine Coverage 85
5.2.8. Controlled and Observed Coverage 88
5.3. Use Model 89
5.3.1. Instrument Code 89
5.3.2. Record Metrics 90
5.3.3. Analyze Measurements 90
5.4. Summary 95

6. Assertion Coverage 97
6.1. What Are Assertions? 97
6.2. Measuring Assertion Coverage 102
6.3. Open Verification Library Coverage 103
6.4. Static Assertion Coverage 104
6.5. Analyzing Assertion Coverage 104
6.5.1. Checker Assertions 105
6.5.2. Coverage Assertions 106
6.6. Summary 107

7. Coverage-Driven Verification . . . 109
7.1. Objections to Coverage-Driven Verification . . . 110
7.2. Stimulus Generation . . . 112
7.2.1. Generation Constraints . . . 113
7.2.2. Coverage-Directed Generation . . . 115
7.3. Response Checking . . . 120
7.4. Coverage Measurement . . . 122
7.4.1. Functional Coverage . . . 123
7.4.2. Code Coverage . . . 124
7.4.3. Assertion Coverage . . . 126
7.4.4. Maximizing Verification Efficiency . . . 127
7.5. Coverage Analysis . . . 129
7.5.1. Generation Feedback . . . 129
7.5.2. Coverage Model Feedback . . . 130
7.5.3. Hole Analysis . . . 131
7.6. Summary . . . 136

8. Improving Coverage Fidelity With Hybrid Models . . . 139
8.1. Sample Hybrid Coverage Model . . . 140
8.2. Coverage Overlap . . . 147
8.3. Static Verification Coverage . . . 149
8.4. Summary . . . 150

Appendix A: *e* Language BNF . . . 151

Index . . . 193

Foreword

As the complexity of today's ASIC and SoC designs continues to increase, the challenge of verifying these designs intensifies at an even greater rate. Advances in this discipline have resulted in many sophisticated tools and approaches that aid engineers in verifying complex designs. However, the age-old question of *when is the verification job done*, remains one of the most difficult questions to answer. And, the process of measuring verification progress is poorly understood.

For example, consider automatic random stimulus generators, model-based test generators, or even general-purpose constraint solvers used by high-level verification languages (such as ***e***). At issue is knowing which portions of a design are repeatedly exercised from the generated stimulus — and which portions of the design are not touched at all. Or, more fundamentally, exactly what functionality has been exercised using these techniques. Historically, answering these questions (particularly for automatically generated stimulus) has been problematic. This challenge has led to the development of various coverage metrics to aid in measuring progress, ranging from *code coverage* (used to identify unexercised lines of code) to contemporary *functional coverage* (used to identify unexercised functionality). Yet, even with the development of various forms of coverage and new tools that support coverage measurement, the use of these metrics within the verification flow tends to be ad-hoc, which is predominately due to the lack of well-defined, coverage-driven verification methodologies.

Prior to introducing a coverage-driven verification methodology, *Functional Verification Coverage Measurement and Analysis* establishes a sound foundation for its readers by reviewing an excellent and comprehensive list of terms that is common to the language of coverage. Building on this knowledge, the author details various forms of measuring progress that have historically been applicable to a traditional verification flow, as well as new forms applicable to a contemporary verification flow.

Functional Verification Coverage Measurement and Analysis is the first book to introduce a useful taxonomy for coverage metric classification. Using this taxonomy, the reader clearly understands the process of creating an effective coverage model. Ultimately, this book presents a coverage-driven verification methodology that integrates multiple forms of coverage and strategies to help answer the question *when is the verification job done*.

Andrew Piziali has created a wonderfully comprehensive textbook on the language, principles, and methods pertaining to the important area of *Functional Verification Coverage Measurement and Analysis*. This book should be a key reference in every engineer's library.

Harry Foster
Chief Methodologist
Jasper Design Automation, Inc.

Andy and I disagree on many fronts: on the role of governments, on which verification language is best, on gun control, on who to work for, on the best place to live and on the value of tightly integrated tools. But, we wholeheartedly agree on the value of coverage and the use of coverage as a primary director of a functional verification process.

Yesterday, I was staring at a map of the Tokyo train and subway system. It was filled with unfamiliar symbols and names — yet eerily similar to maps of other subway systems I am more familiar with. Without a list of places I wished to see, I could wander for days throughout the city, never sure that I was visiting the most interesting sites and unable to appreciate the significance of the sites that I was haphazardly visiting. I was thus armed with a guide book and recommendations from past visitors. By constantly checking the names of the stations against the stations on my intended route, I made sure I was always traveling in the correct direction, using the shortest path. I was able to make the most of my short stay.

Your next verification project is similar: it feels familiar — yet it is filled with new features and strange interactions. A verification plan is necessary to identify those features and interactions that are the most important. The next step, using coverage to measure your progress toward that plan, is just as crucial. Without it, you may be spending your effort in redundant activities. You may also not realize that a feature or interaction you thought was verified was, in fact, left completely unexercised. A verification plan and coverage metrics are essential tools in ensuring that you make the most of your verification resources.

This book helps transform the art of verification planning and coverage measurement into a process. I am sure it will become an important part of the canons of functional verification.

Janick Bergeron
Scientist
Synopsys
Tokyo, April 2004

Preface

Functional verification is consuming an ever increasing share of the effort required to design digital logic devices. At the same time, the cost of bug escapes[1] and crippled feature sets is also rising as missed market windows and escalating mask set costs take their toll. Bug escapes have a number of causes but one of the most common is uncertainty in knowing when verification is complete. This book addresses that problem.

There are several good books[2] [3] on the subject of functional verification.[4] However, the specific topic of measuring verification progress and determining when verification is done remains poorly understood. The purpose of this book is to illuminate this subject. The book is organized as follows.

The introduction chapter is an overview of the general verification problem and the methods employed to solve it.

Chapter 1," The Language of Design Verification," defines the terminology I use throughout the book, highlighting the nuances of similar terms.

Chapter 2, "Functional Verification," defines functional verification, distinguishes it from test and elaborates the functional verification process.

Chapter 3, "Measuring Verification Coverage," introduces the basics of coverage measurement and analysis: coverage metrics and coverage spaces.

---

[1] Logic design bugs undetected in pre-silicon verification.

[2] Writing Testbenches, Second Edition, Janick Bergeron, Kluwer Academic Publishers, 2003

[3] Assertion-Based Design, Harry D. Foster, Adam C. Krolnik, David J. Lacey, Kluwer Academic Publishers, 2003

[4] "Design verification" and "functional verification" are used interchangeably throughout this book.

Chapter 4, "Functional Coverage," delves into coverage derived from specifications and the steps required to model the design intent derived from the specifications. Two specific kinds of functional coverage are also investigated: temporal coverage and finite state machine (FSM) coverage.

Chapter 5, "Code Coverage," explains coverage derived from the device implementation, the RTL. It addresses the various structural and syntactic RTL metrics and how to interpret reported data.

Chapter 6, "Assertion Coverage," first answers the question of "Why would I want to measure coverage of assertions?" and then goes on to describe how to do so.

Chapter 7, "Coverage-Driven Verification," integrates all of the previous chapters to present a methodology for minimizing verification risk and maximizing the rate at which design bugs are exposed. In this chapter, I explain stimulus generation, response checking and coverage measurement using an autonomous verification environment. The interpretation and analysis of coverage measurements and strategies for reaching functional closure — i.e. 100% coverage — are explained.

Chapter 8, "Improving Coverage Fidelity with Hybrid Models," introduces the concept of coverage model fidelity and the role it plays in the coverage process. It suggests a means of improving coverage fidelity by integrating coverage measurements from functional, code and assertion coverage into a heterogeneous coverage model.

The Audience

There are two audiences to which this book addressed. The first is the student of electrical engineering, majoring in digital logic design and verification. The second is the practicing design verification — or hardware design — engineer.

When I was a student in electrical engineering (1979), no courses in design verification were offered. There were two reasons for this. The first was that academia was generally unaware of the magnitude of the verification challenge faced by logic designers of the most complex designs: mainframes and supercomputers. Second, no textbooks were available on the subject. Both of these reasons have now been dispensed with so this book may be used in an advanced design verification course.

The practicing design verification and design engineer will find this book useful for becoming familiar with coverage measurement and analysis.

It will also serve as a reference for those developing and deploying coverage models.

Prerequisites

The reader is expected to have a basic understanding of digital logic design, logic simulation and computer programming.

Acknowledgements

I want to thank my wife Debbie and son Vincent for the solitude they offered me from our limited family time. My technical reviewers Mark Strickland, Shmuel Ur, Mike Kantrowitz, Cristian Amitroaie, Mike Pedneau, Frank Armbruster, Marshall Martin, Avi Ziv Harry Foster, Janick Bergeron, Shlomi Uziel, Yoav Hollander, Ziv Binyamini and Jon Shiell provided invaluable guidance and feedback from a variety of perspectives I lack. Larry Lapides kept my pride in writing ability in check with grammar and editing corrections. My mentors Tom Kenville and Vern Johnson pointed me in the direction of "diagnostics development," later known as design verification. The Unix text processing tool suite groff and its siblings — the -ms macros, gtbl, geqn and gpic — allowed me to write this book using my familiar Vim text editor and decouple typesetting from the composition process, as it should be. Lastly, one of the most fertile environments for innovation, in which my first concepts of coverage measurement were conceived, was enriched by Convex Computer colleagues Russ Donnan and Adam Krolnik.

Introduction

What is functional verification? I introduce a formal definition for functional verification in the next chapter, "The Language of Design Verification," and explore it in depth in chapter 2, "Functional Verification." For now, let's just consider it the means by which we discover functional logic errors in a representation of the design, whether it be a behavioral model, a register transfer level (RTL) model, a gate level model or a switch level model. I am going to refer to any such representation as "the device" or "the device-under-verification" (DUV). Functional verification is not timing verification or any other back-end validation process.

Logic errors (bugs) are discrepancies between the intended behavior of the device and its observed behavior. These errors are introduced by the designer because of an ambiguous specification, misinterpretation of the specification or a typographical error during model coding. The errors vary in abstraction level depending upon the cause of the error and the model level in which they were introduced. For example, an error caused by a specification misinterpretation and introduced into a behavioral model may be algorithmic in nature while an error caused by a typo in the RTL may topological. How do we expose the variety of bugs in the design? By verifying it! The device may be verified using static, dynamic or hybrid methods. Each class is described in the following sections.

Static Methods

The static verification methods are model checking, theorem proving and equivalence checking.

Model checking demonstrates that user-defined properties are never violated for all possible sequences of inputs.

Theorem proving demonstrates that a theorem is proved — or cannot be proved — with the assistance of a proof engine.

Equivalence checking, as its name implies, compares two models against one another to determine whether or not they are logically equivalent. The models are not necessarily at the same abstraction level: one may be RTL while the other is gate level. Logical equivalence means two circuits implement the same Boolean logic function, ignoring latches and registers.

There are two kinds of equivalence checking: combinational and sequential.[1] Combinational equivalence checking uses a priori structural information found between latches. Sequential equivalence checking detects and uses structural similarities during state exploration in order to determine logical equivalence across latch boundaries.

Lastly, I should mention that Boolean satisfiability (SAT) solvers are being employed more frequently for model checking, theorem proving and equivalence checking. These solvers find solutions to Boolean formulae used in these static verification techniques.

Dynamic Methods

A dynamic verification method is characterized by simulating the device in order to stimulate it, comparing its response to the applied stimuli against an expected response and recording coverage metrics. By "simulating the device," I mean that an executable software model — written in a hardware description language — is executed along with a verification environment. The verification environment presents to the device an abstraction of its operational environment, although it usually exaggerates stimuli parameters in order to stress the device. The verification environment also records verification progress using a variety of coverage measurements discussed in this book.

Static versus Dynamic Trade-offs

The trade-off between static and dynamic method is between capacity and completeness. All static verification methods are hampered by capacity constraints that limit their application to small functional blocks of a device At the same time, static methods yield a complete, comprehensive verification of the proven property. Together, this leads to the application of static methods to small, complex logic units such as arbiters and bus controllers.

---

[1] C.A.J. van Eijk, "Sequential Equivalence Checking Based on Structural Similarities," IEEE Trans. CAD of ICS, July 2000.

Dynamic methods, on the other hand, suffer essentially no capacity limitations. The simulation rate may slow dramatically running a full model of the device, but it will not fail. However, dynamic methods cannot yield a complete verification solution because they do not perform a proof.

There are many functional requirements whose search spaces are beyond the ability to simulate in a lifetime. This is because exhaustively exercising even a modest size a device may require an exorbitant number of simulation vectors. If a device has N inputs and M flip-flops, $(2^N)^M$ stimulus vectors may be required[2] to fully exercise it. A modest size device may have 10 inputs and 100 flip-flops (just over three 32-bit registers). This device would require $(2^{10})^{100}$, or $2^{1,000}$ vectors to fully exercise. If we were to simulate this device at 1,000 vectors per second, it would take 339,540,588,380, 062,907,492,466,172,668,391,072,376,037,725,725,208,993,588,689,808, 600,264,389,893,757,743,339,953,988,988,382,771,724,040,525,133,303, 203,524,078,771,892,395,266,266,335,942,544,299,458,056,845,215,567, 848,460,205,301,551,551,163,124,606,262,994,092,425,972,759,467,835, 103,001,336,336,717,048,865,167,147,297,613,428,902,897,465,679,093, 821,821, 978, 784, 398, 755, 534, 655, 038, 141, 450, 059, 156, 501 years[3] to exhaustively exercise. Functional requirements that must be exhaustively verified should be proved through formal methods.

Hybrid Methods

Hybrid methods, also known as semi-formal methods, combine static and dynamic techniques in order to overcome the capacity constraints imposed by static methods alone while addressing the inherent completeness limitations of dynamic methods. This is illustrated with two examples.

Suppose we postulate a rare, cycle distant[4] device state to be explored by simulating forward from that state. The validity of this device state may be proven using a bounded model checker. The full set of device properties may be proven for this state. If a property is violated, the model checker will provide a counter example from which we may deduce a corrective

[2] I say "may be required" because it depends upon the complexity of the device. If the device simply latches its N inputs into $\frac{M}{N}$-deep FIFOs, it would only require $\frac{M}{N}$ vectors to exhaustively exercise.

[3] Approximately 3.4×10^{308} years.

[4] "Distant" in the sense that it is many, many cycles from the reset state of the device, perhaps too many cycles to reach in practice.

modification to the state. Once the state is fully specified, the device may be placed in the state using the simulator's broadside load capability. Simulation may then *start* from this point, as if we had simulated to it from reset.

The reverse application of static and dynamic methods may also be employed. Perhaps we discovered an unforeseen or rare device state while running an interactive simulation and we are concerned that a device requirement, captured as a property, may be violated. At the simulation cycle of interest, the state of the device and its inputs are captured and specified as the initial search state to a model checker. The model checker is then directed to prove the property of concern. If the property is violated, any simulation sequence that reached this state is a counter-example.

Summary

In this introduction, I surveyed the most common means of functionally verifying a design: static methods, dynamic methods and hybrid methods. In the next chapter, The Language of Coverage, I methodically define the terminology used throughout the remainder of the book.

1. The Language of Coverage

Stripped of all of its details, design verification is a communication problem. Ambiguities lead to misinterpretations which lead to design errors. In order to clearly convey the subject of coverage measurement and analysis to you, the reader, we must communicate using a common language. In this chapter, I define the terminology used throughout the rest of the book. It should be referenced whenever an unfamiliar word or phrase is encountered.

You will find references to the high-level verification language ***e*** in this glossary. I use ***e*** to illustrate the implementation of coverage models in this book. The ***e*** language syntax may be referenced in appendix A. You may find the complete language definition in the "***e*** Language Reference Manual," available at the IEEE 1647 web site, `http://www.ieee1647.org/`.

assertion	An expression stating a safety (invariant) or liveness (eventuality) property.
assertion coverage	The fraction of device assertions executed and passed or failed. Assertion coverage is the subject of chapter 6.
assertion coverage density	The number of assertions evaluated per simulation cycle.
attribute	In the context of the device, a parameter or characteristic of an input or output on an interface. In the context of a coverage model, a parameter or dimension of the model. Attributes and their application is discussed in chapter 4, "Functional Coverage."

branch coverage	A record of executed, alternate control flow paths, such as those through an if-then-else statement or case statement. Branch coverage is the subject of section 5.2.3.
checker coverage	The fraction of verification environment checkers executed and passed or failed.
code coverage	A set of metrics at the behavioral or RTL abstraction level which define the extent to which the design has been exercised. Code coverage is the subject of chapter 5.
code coverage density	The number of code coverage metrics executed or evaluated per simulation cycle. A metric may be a line, statement, branch, condition, event, bit toggle, FSM state visited or FSM arc traversed.
condition coverage	A record of Boolean expressions and subexpressions executed, usually in the RTL. Also known as *expression coverage*. Condition coverage is discussed in section 5.2.4.
coverage	A measure of verification completeness.
coverage analysis	The process of reviewing and analyzing coverage measurements. Coverage analysis is discussed in section 7.5.
coverage closure	Reaching a defined coverage goal.
coverage database	A repository of recorded coverage observations. For code coverage, counts of observed metrics such as statements and expressions may be recorded. For functional coverage, counts of observed coverage points are recorded.

coverage density	The number of coverage metrics observed per simulation cycle. See also *functional coverage density*, *code coverage density* and *assertion coverage density*.
coverage goal	That fraction of the aggregate coverage which must be achieved for a specified design stage, such as unit level integration, cluster integration and functional design freeze.
coverage group	A related set of attributes, grouped together for implementation purposes at a common correlation time. In the context of the ***e*** language, a struct member defining a set of items for which data is recorded.
coverage item	The implementation level parallel to an attribute. In the context of the ***e*** language, a coverage group member defining an attribute.
coverage measurement	The process of recording points within a coverage space.
coverage metric	An attribute to be used as a unit of measure and recorded, which defines a dimension of a coverage space. The role of coverage metrics is the subject of chapter 3, "Measuring Verification Coverage."
coverage model	An abstract representation of device behavior composed of attributes and their relationships. Coverage model design is discussed in chapter 4, "Functional Coverage."
coverage point	A point within a multi-dimensional coverage model, defined by the values of its attributes.

coverage report — A summary of the state of verification progress — as measured by coverage — capturing all facets of coverage at multiple abstraction levels.

coverage space — A multi-dimension region defined by the attributes of the coverage space and their values. Usually synonymous with "coverage model." The following diagram illustrates a coverage space.

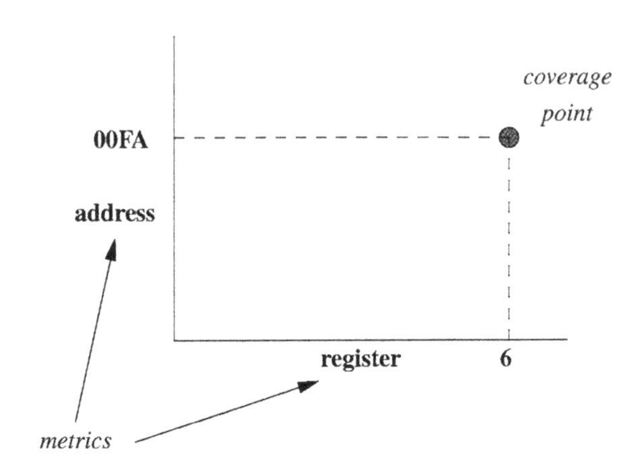

The coverage space is discussed in section 3.2.

cross coverage — A coverage model whose space is defined by the full permutation of all values of all attributes. More precisely known as multi-dimensional matrix coverage. Cross coverage is discussed in section 4.3.2, "Attribute Relationships."

data coverage — Coverage measurements in the data domain of the device behavior.

device — Device to be verified. Sometimes referred to as the device-under-verification (DUV).

DUT — Acronym for "device under test;" i.e. the device to be tested. This is distinguished from DUV (device under verification) in that a DUV is verified while a DUT is tested.

DUV — Acronym for "device under verification;" i.e. the device to be verified. This is distinguished from DUT (device under test) in that a DUT is tested while a DUV is verified.

e — A high-level verification language (HLVL) invented by Yoav Hollander and promoted by Verisity Design. The BNF of the ***e*** language is in appendix A. The "***e*** Language Reference Manual" may be referenced from `http://www.ieee1647.org/`.

event — Something which defines a moment in time such as a statement executing or a value changing. In the context of the ***e*** language, a struct member defining a moment in time. An ***e*** event is either explicitly *emitted* using the *emit* action or implicitly *emitted* when its associated temporal expression succeeds.

explicit coverage — Coverage whose attributes are explicitly chosen by the engineer rather than being a characteristic of the measurement interface.

expression coverage — A record of Boolean expressions and subexpressions executed, usually in the RTL. Also known as *condition coverage*. Expression coverage is discussed in section 5.2.4, "Condition Coverage."

fill — To fill a coverage space means to reach the coverage goal of each point within that space.

functional coverage	Coverage whose metrics are derived from a functional or design specification. Functional coverage is the subject of chapter 4.
functional coverage density	The number of functional coverage points traversed per simulation cycle. Coverage density is discussed in section 7.4.4, “Maximizing Verification Efficiency.”
grade	For a single coverage model, the fraction of the coverage space it defines which has been observed. Regions of the coverage space or individual points may be unequally weighted. For a set of coverage models, a weighted average of the grade of each model.
hit	Observing a defined coverage point during a simulation.
HLVL	High-level verification language. A programming language endowed with semantics specific to design verification such as data generation, temporal evaluation and coverage measurement.
hole	A defined coverage point which has not yet been observed in a simulation or a set of such points sharing a common attribute or semantic.
implicit coverage	Coverage whose attributes are implied by characteristics of the measurement interface rather than explicitly chosen by the engineer.
input coverage	Coverage measured at the primary inputs of a device.
internal coverage	Coverage measured on an internal interface of a device.

line coverage	The fraction of RTL source lines executed by one or more simulations. Line coverage is discussed in section 5.2.1, "Line Coverage."
merge coverage	To coalesce the coverage databases from a number of simulations.
model	An abstraction or approximation of a logic design or its behavior.
output coverage	Coverage measured at the primary outputs of a device.
path coverage	The fraction of all control flow paths executed during one or more simulations. Path coverage is discussed in section 5.2.3, "Branch Coverage."
sample	To record the value of an attribute.
sampling event	A point in time at which the value of an attribute is sampled. Sampling time is discussed in section 4.3.1, "Attribute Identification."
sequential coverage	A composition of data and temporal coverage wherein specific data patterns applied in specific sequences are recorded.
statement coverage	The fraction of all language statements — behavioral, RTL or verification environment — executed during one or more simulations. See section 5.2.2 for an example of statement coverage.
temporal	Related to the time domain behavior of a device or its verification environment.

temporal coverage	Measurements in the time domain of the behavior of the device.
test	The verb "test" means executing a series of trials on the device to determine whether or not its behavior conforms with its specifications. The noun "test" refers to either a trial on the device or to the stimulus applied during a specific trial. If referring to stimulus, it may also perform response checking against expected results.
toggle coverage	A coverage model in which the change in value of a binary attribute is recorded. Toggle coverage is discussed in section 5.2.6.
verification	The process of demonstrating the intent of a design is preserved in its implementation.
verification interface	An abstraction level at which a verification process is performed. If dynamic verification (simulation) is used, this is a common interface at which stimuli are applied, behavioral response is checked and coverage is measured.
verify	Demonstrate the intent of a design is preserved in its implementation.
weight	A scaling factor applied to an attribute when calculating cumulative coverage of a single coverage model or applied to a coverage model when totaling cumulative coverage of all coverage models.

weighted average — The sum of the products of fractional coverage times weight, divided by the sum of their weights.

I.e. $$weighted\ average = \frac{\sum_{i=1}^{N} coverage_i * weight_i}{\sum_{i=1}^{N} weight_i},$$

where $coverage_i$ is a particular coverage measurement, $weight_i$ is the weight of the measurement and N is the number of coverage models.

2. Functional Verification

In this chapter, I define functional verification, distinguish verification from testing and outline the functional verification process.

What is functional verification? A definition which has served me well for many years is the following: *Functional verification is demonstrating the intent of a design is preserved in its implementation.* In order to thoroughly understand functional verification, we need to understand this definition. The following diagram[1] is useful for explaining the definition.

[1] Tony Wilcox, personal whiteboard discussion, 2001.

2.1. Design Intent Diagram

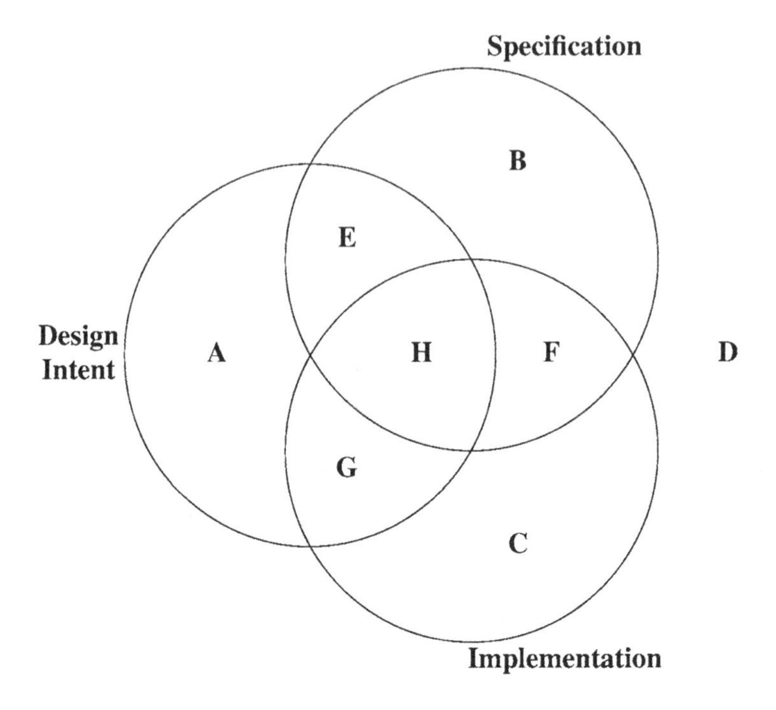

Figure 2-1 Design Intent Diagram

The diagram is composed of three overlapping circles, labeled "Design Intent," "Specification" and "Implementation." All areas in the diagram represent device behavior. The space defined by the union of all of the regions (A through G) represents the potential behavioral space of a device. The region outside the three circles, D, represents unintended, unspecified and unimplemented behavior. The first circle, "Design Intent" ($A \cup E \cup G \cup H$),[2] represents the intended behavior of the device, as conceived in the mind's eye(s) of the designer(s). The second circle, "Specification" ($B \cup E \cup F \cup H$), bounds the intent captured by the device functional specification. The third circle, "Implementation" ($C \cup F \cup G \cup H$), captures the design intent implemented in the RTL.

[2] The conventional set operators are used. $\cup$ for set union, $\cap$ for set intersection, $\subseteq$ for subset, $\subset$ for proper subset and – for set exclusion.

If the three circles were coincident, i.e. region H defined all three circles, all intended device behavior would be specified and captured in the device implementation, but no more. However, in reality, this is rarely the case. Let's examine the remaining regions to understand why this is so.

Region E is design intent captured in the specification but absent from the implementation. Region F is unintended behavior which is nonetheless specified and implemented (!). Region G is implemented, intended behavior which was not captured in the specification.

Region EH ($E \cup H$) represents design intent successfully captured by the specification but only partially implemented. The remaining part of the specification space, BF ($B \cup F$), is unintended yet specified behavior. This is usually results from gratuitous embellishment or feature creep.

Region FH ($F \cup H$) represents specified behavior successfully captured in the implementation. The remaining part of the implementation space, CG ($C \cup G$), is unspecified yet implemented behavior. This could also be due to gratuitous embellishment or feature creep. Region GH ($G \cup H$) represents intended and implemented behavior.

There are four remaining regions to examine. The first, AE ($A \cup E$), is unimplemented yet intended behavior. The second, AG ($A \cup G$), is unspecified yet intended behavior. The third, BE ($B \cup E$), is specified yet unimplemented behavior. The fourth, CF ($C \cup F$), is unintended yet implemented behavior.

The objective of functional verification is to bring the device behavior represented by each of the three circles — design intent, specification and implementation — into coincidence. To do so, we need to understand the meaning of design intent, where it comes from and how it is transformed in the context of functional verification.

2.2. Functional Verification

A digital logic design begins in the mind's eye of the system architect(s). This is the original intent of the design, its intended behavior. From the mind, it goes through many iterations of stepwise refinement until the layout file is ready for delivery to the foundry. Functional verification is an application of information theory, supplying the redundancy and error-correcting codes required to preserve information integrity through the design cycle. The redundancy is captured in natural (human) language specifications.

However, there are two problems with this explanation. First of all, this "original intent" is incomplete and its genesis is at a high abstraction level. The concept for a product usually begins with a marketing requirements document delivered to engineering. An engineering system architect invents a product solution for these requirements, refining the abstract requirements document into a functional specification. The design team derives a design specification from the functional specification as they specify a particular microarchitectural implementation of the functionality.

The second problem with the explanation is that, unlike traditional applications of information theory, where the message should be preserved as it is transmitted through the communication channel, it is intentionally refined and becomes less abstract with each transformation through the design process. Another way to look at the design process is that the message is incrementally refined, clarified and injected into the communication channel at each stage of design. Next, let's distinguish implementation from intent.

In this context, the implementation is the RTL (Verilog, SystemVerilog or VHDL) realization of the design. It differs from intent in that it is not written in a natural language but in a rigorous machine readable language. This removes both ambiguity and redundancy, allowing a logic compiler to translate the code into a gate description, usually preserving the semantics of the RTL. Finally, what is meant by "demonstrate" when we write "demonstrate the intent of a design is preserved in its implementation?"

Verification, by its very nature, is a comparative process. This was not apparent to a director of engineering I once worked for. When I insisted his design team update the design specification for the device my team was verifying, he replied: "Andy, the ISP *is* the specification!" (ISP was a late eighties hardware design language.) That makes one's job as a verification engineer quite easy, doesn't it? By definition, that design was correct as written because the intent — captured in an ISP "specification" — and implementation were claimed to be one and the same. The reality was the system architect and designers held the design intent in their minds but were unwilling to reveal it in an up-to-date specification for use by the verification team.

The intent of a design is demonstrated to have been preserved through static and dynamic methods. We are concerned with dynamic methods in this book, executing the device in simulation in order to observe and compare its behavior against expected behavior. Now, let's look at the difference between testing and verification.

2.3. Testing versus Verification

Many engineers mistakenly use the terms "test" and "verification" interchangeably. However, testing is but one way to verify a design, and a less rigorous and quantitative approach at that. Why is that?

Writing for "Integrated System Design" in 2000, Gary Smith wrote: "The difference between 'test' and 'verification' is often overlooked ... You test the device ... to insure that the implementation works. ... Verification ... checks to see if the hardware or software meets the requirements of the original specification."[3] There are subtle, but important, differences between the two.

Testing is the application of a series of tests to the DUT[4] to determine if its behavior, for each test, conforms with its specifications. It is a sampling process to assess whether or not the device works. A sampling process? Yes. It is a sampling process because not all aspects of the device are exercised. A subset of the totality of possible behaviors is put to the test.

A test also refers to the stimulus applied to the device for a particular simulation and may perform response checking against expected behavior. Usually, the only quantitative measure of progress when testing is the number of tests written and the number of tests passed although, in some instances, coverage may also be measured. Hence, it is difficult to answer the question "Have I explored the full design space?"

Verification encompasses a broad spectrum of approaches to discovering functional device flaws. In this book, we are concerned with those approaches which employ coverage to measure verification progress. Let us examine an effective verification process.

2.4. Functional Verification Process

The functional verification process begins with writing a verification plan, followed by implementing the verification environment, device bring-up and device regression. Each of these steps is discussed in the following

[3] Gary Smith, "The Dream," "Integrated System Design," December 2000

[4] See definitions of test, verify, DUT and DUV in chapter 1, "The Language of Coverage."

sections.

2.4.1. Functional Verification Plan

The verification plan defines what must be verified and how it will be verified. It describes the scope of the verification problem for the device and serves as the functional specification for the verification environment. Dynamic verification (i.e. simulation-based) is composed of three aspects, as illustrated below in figure 2-2.

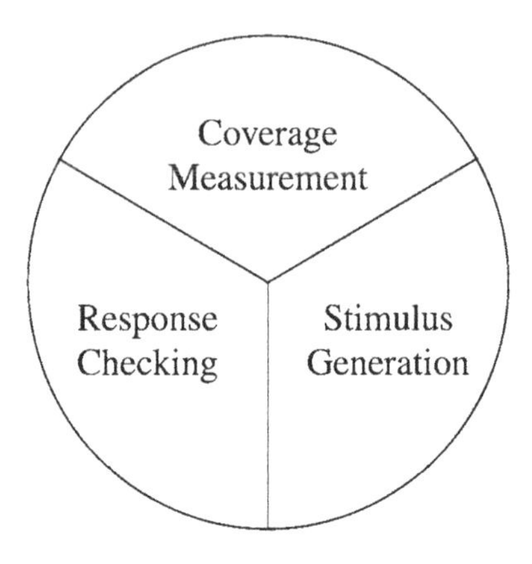

Figure 2-2 Functional Verification Aspects

This leads to one of three orthogonal partitions of the verification plan: first, by verification aspect. The scope of the verification problem is defined by the coverage section of the verification plan. The stimulus generation section defines the machinery required to generate the stimuli required by the coverage section. The response checking section describes the mechanisms to be used to compare the response of the device to the expected, specified response.

The second partitioning of the verification plan is between verification requirements derived from the device functional specification and those derived from its design specification. These are sometimes called architecture and implementation verification, as illustrated below in figure 2-3.

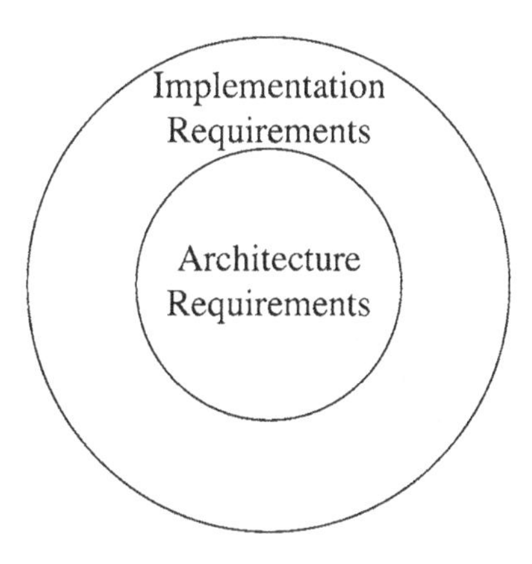

Figure 2-3 Functional Verification Requirements Sources

Architecture verification concerns itself with exposing device behaviors which deviate from its functional behavioral requirements. For example, if an add instruction is supposed to set the overflow flag when the addition results in a carry out in the sum, this is an architectural requirement. Implementation verification is responsible for detecting deviations from microarchitectural requirements specified by the design specification. An example of an implementation requirement is that a read-after-write register dependency in an instruction pair must cause the second instruction to read from the register bypass rather than the register file.

The third partitioning of the verification plan is between *what* must be verified and *how* it is to be verified. The former draws its requirements from the device functional and design specifications while the latter captures the top-level and detailed design of the verification environment itself. What must be verified is captured in the functional, code and assertion coverage requirements of the coverage measurement section of the verification plan. How the device is to be verified is captured in the top- and detailed-design section of each of the three aspects of the verification plan: coverage measurement, stimulus generation and response checking.

In the following three sections, we examine each of the verification aspects in more detail.

2.4.1.1. Coverage Measurement

The coverage measurement section of the verification plan — sometimes referred to as the coverage plan — should describe the extent of the verification problem and how it is partitioned, as discussed above. It should delegate responsibility for measuring verification progress among the kinds of coverage and their compositions: functional, code, assertion and hybrid.[5] The functional coverage section of the coverage plan should include the top-level and detailed design of each of the coverage models. The code coverage section should specify metrics to be employed, coverage goals and gating events for deploying code coverage. For example, you should be nearing full functional coverage and have stable RTL before turning on code coverage measurement. The responsibility of assertion coverage in your verification flow should also be discussed.

Next, we need to consider how stimulus will be generated to achieve full coverage.

2.4.1.2. Stimulus Generation

The stimulus required to fully exercise the device — that is, to cause it to exhibit all possible behaviors — is the responsibility of the stimulus generation aspect of the verification environment. Historically, a hand-written file of binary vectors, one vector (line) per cycle, served as simulation stimulus. In time, symbolic representations of vectors such as assembly language instructions were introduced, along with procedural calls to vector generation routines. Later, vector generators were developed, beginning with random test program generators (RTPG)[6] and evolving through model-based test generators (MBTG)[7] to the general purpose constraint solvers of current high-

---

[5] These coverage techniques are described in chapters 4, 5 and 6: "Functional Coverage,", "Code Coverage" and "Assertion Coverage." The application of these coverage techniques is explained in chapter 7, "Coverage-Driven Verification" while their composition is the subject of chapter 8, "Improving Coverage Fidelity With Hybrid Models."

[6] Reference the seminal paper "Verification of the IBM RISC System/6000 by a Dynamic Biased Pseudo-Random Test Program Generator" by A. Aharon, A. Ba-David, B. Dorfman, E. Gofman, M. Leibowitz, V. Schwartzburd, IBM Systems Journal, Vol. 30, No. 4, 1991.

[7] See "Model-Based Test Generation for Processor Design Verification" by Y.

level verification languages (HLVL).[8]

In this book, I illustrate verification environment implementations using the HLVL ***e***. As such, the stimulus generation aspect is composed of generation constraints and sequences. Generation constraints are statically declared rules governing data generation. Sequences define a mechanism for sending coordinated data streams or applying coordinated actions to the device.

Generation constraints are divided into two sets according to their source: the functional specification of the device and the verification plan. The first set of constraints are referred to as functional constraints because they restrict the generated stimuli to valid stimuli. The second set of constraints are known as verification constraints because they further restrict the generated stimuli to the subset useful for verification. Let's briefly examine each constraint set.

Although there are circumstances in which we may want to apply invalid stimulus to the device, such as verifying error detection logic, in general only valid stimuli are useful. Valid stimuli are bounded by both data and temporal rules. For example, if we are generating instructions which have an opcode field, its functional constraint is derived from the specification of the opcode field. This specification should be referenced by the stimulus section of the verification plan. If we are generating packet requests whose protocol requires a one cycle delay between grant and the assertion of valid, the verification plan should reference this temporal requirement.

In addition to functional constraints, verification constraints are required to prune the space of all valid stimuli to those which exercise the device boundary conditions. What are boundary conditions? Depending upon the abstraction level — specification or implementation — a boundary condition is either a particular situation described by a specification or a condition for which specific logic has been created. For example, if the specification says that when a subtract instruction immediately follows an add instruction and both reference the same operand, the ADDSUB performance monitoring flag is set, this condition is a boundary condition. Functional and verification constraints are discussed further in the context of coverage-driven verification in chapter 7.

Lichtenstein, Y. Malka and A. Aharon, Innovative Applications of Artificial Intelligence, AAAI Press, 1994.

[8] Reference U.S. patent 6,219,809, "System and Method for Applying Flexible Constraints," Amos Noy (Verisity Ltd.), April 17, 2001

Having designed the machinery required to generate the stimulus required to reach 100% coverage, how do we know that the device is behaving properly during each of the simulations? This is the subject of the next section of this chapter, response checking.

2.4.1.3. Response Checking

The response checking section of the verification plan is responsible for describing how the behavior of the device will be demonstrated to conform with its specifications. There are two general strategies employed: a reference model or distributed data and temporal checks.

The reference model approach requires an implementation of the device at an abstraction level suitable for functional verification. The abstraction level is apparent in each of the device specifications: the functional specification and the design specification. The functional specification typically describes device behavior from a black box[9] perspective. The design specification addresses implementation structure, key signals and timing details such as pipelining. As such, a reference model should only be used for architecture verification, not implementation verification. If used for implementation verification, such a model would result in a second implementation of the device at nearly the same abstraction level as the device itself. This would entail substantial implementation and maintenance costs because the model would have to continually track design specification changes, which are often quite frequent.

A consideration for choosing to use a reference model is that the reference model must itself be verified. Will this verification entail its own, recursive process? Although any device error reported by a verification environment must be narrowed down to either an error in the DUV or an error in the verification environment, the effort required to verify a complex reference model may be comparable to verifying the device itself.

Another consideration when using a reference model is that it inherently exhibits implementation artifacts. An implementation artifact is an unspecified behavior of the model resulting from implementation choices. This unspecified behavior must not be compared against the device behavior because it is not a requirement.

---

[9] Why aren't black box and white box verification called opaque and transparent box verification? After all, black box verification means observing device behavior from its primary I/Os alone. White box verification means observing internal signals and structures.

In the context of response checking using a reference model, the *executable specification* is often cited. *Executable specification* is really an oxymoron because a specification should only define device requirements at some abstraction level. An executable model, on the other hand, must be defined to a sufficient level of detail to be run by a computing system. The implementation choices made by the model developer invariably manifest themselves as behavioral artifacts not required by the device specification. In other words, the device requirements and model artifacts are indistinguishable from one another.

The second response checking strategy, distributed checks, uses data and temporal monitors to capture device behavior. This behavior is then compared against expected behavior. One approach used for distributed checking is the scoreboard.

A scoreboard is a data structure used to store either expected results or data input to the device. Consider the device and scoreboard illustrated below in figure 2-4.

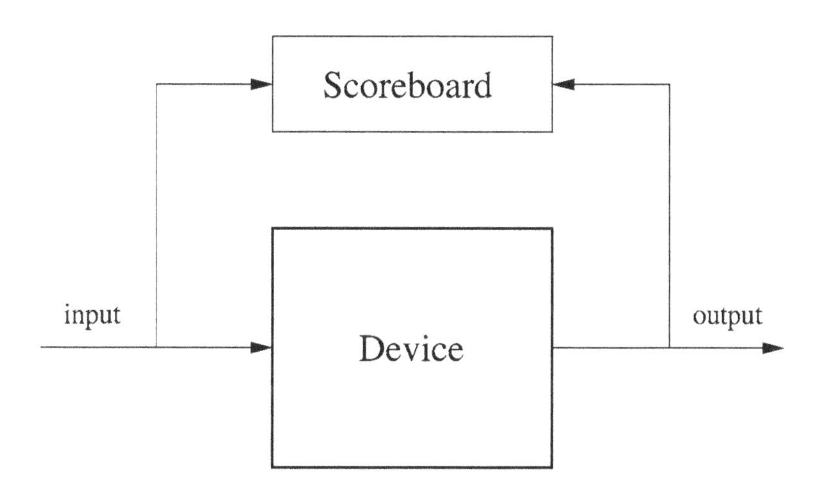

Figure 2-4 Scoreboard

The device captures input on the left, transforms it, and writes output on the right. Consider the case where packets are processed by the device. If expected results are stored in the scoreboard, each packet is transformed per the device specification before being written to the scoreboard. Each time a packet is output by the device, it is compared with the transformed packet in the scoreboard and that packet is removed from the scoreboard.

If input data is stored in the scoreboard, each packet read by the device is also written to the scoreboard. Each time a packet is output by the device, the associated input packet is read from the scoreboard, transformed according to the specification, and compared with the output packet.

Reference models and distributed checks are often employed at the same time in a heterogeneous fashion. For example, an architectural reference model may be used for processor instruction set architecture (ISA) checking but distributed checks used for bus protocol checking.

Once the verification plan is written, it must be implemented. We address implementing the verification environment in the next section.

2.4.2. Verification Environment Implementation

The verification plan is written (at least the first draft) and now it is time to turn to use it as a design specification. It will serve as the functional specification for the verification environment. Each of the aspects of the verification environment — coverage, stimulus and checking — are defined in the verification plan. The architecture of each should be partly dictated by reuse.

For the same reasons reusable design IP has become critical for bringing chips to market on time, reusable verification IP has also become important. This has two consequences. First, acquire verification components rather than build them whenever possible. Commercial verification components are available for a variety of common hardware interfaces. You may also find pre-built verification components within your organization. Second, write reusable verification components whenever you build a verification environment.

When ***e*** is used to implement verification IP, the ***e*** Reuse Methodology (***e***RM)[10] should be followed. "The ***e*** Reuse Methodology Developer Manual ... is all about maximizing reusability of verification code written in ***e***. ***e***RM ensures reusability by delivering the best known methods for designing, coding, and packaging ***e*** code as reusable components."[11]

The first application of the verification environment is getting the DUV to take its first few baby steps. This is known as "bring-up," the subject of

---

[10] http://www.verisity.com/.

[11] "***e*** Reuse Methodology Developer Manual," Verisity Design, © 2002-2004.

the next section.

2.4.3. Device Bring-up

The purpose of bring-up is to shake out the blatant, show-stopper bugs that prevent the device from even running. Although the designer usually checks out the basic functionality during RTL development, invariably the first time the verification engineer gets their hands on the block, all hell breaks loose. In order to ease the transition from the incubation period, where the designer is writing code, to hatching, the verification engineer prepares a small set of simulations (also known as *sims*) to demonstrate basic functionality.

These simulations exercise an extremely narrow path through the behavioral space of the device. Each must be quite restricted in order to make it easy to diagnose a failure. Almost anything may fail when the device is first simulated.[12] The process of bringing up a device is composed of making an assumption, demonstrating that the assumption is true and then using the demonstrated assumption to making a more complex assumption.

For example, the first bring-up sim for a processor would simply assert and de-assert the reset pin. If an instruction is fetched from the reset vector of the address space, the first step of functionality is demonstrated. For a packet processing device, the first sim may inject one packet and make sure the packet is routed and reaches egress successfully. If the packet is output, transformed as (or if) required, the simulation passes.

If we are using an autonomous verification environment,[13] rather than directed tests, how do we run such basic simulations? In an aspect-oriented HLVL like ***e***, a constraint file is loaded on top of the environment. This constraint file restricts the generated input to a directed sequence of stimuli. In the case of the processor example above, the constraint file might look like this:

[12] The DV engineer's motto is "If it has not been verified, it doesn't work."

[13] An *autonomous verification environment* is self-directed and composed of stimulus generation, response checking and coverage measurement aspects. It is normally implemented using an HLVL because of its inherent verification semantics.

```
extend reset_s {
  keep all of {
    cycle in [0..3] => reset_sig = 1;
    cycle in [4..5] => reset_sig = 0;
    cycle >= 6      => reset_sig = 1;
    sim_length == 7
  }
}
```

Reset is active low and the field `reset_sig` is written to the processor reset pin. `reset_sig` is constrained to the value one (de-asserted) between cycles zero to three. It is constrained to zero (asserted) for two cycles, starting on cycle four. For all cycles beyond cycle five, it is again constrained to one. Finally, `sim_length` (the number of simulation cycles to run) is constrained to seven.

After the device is able to process basic requests, the restrictions imposed on the verification environment are incrementally loosened. The variety of stimuli are broadened in both the data domain and the time domain. The data domain refers to the spectrum and permutation of data values that may be driven into the device. The time domain refers to the scope of temporal behavior exhibited on the device inputs by the verification environment.

Once the device can run any set of simulations which deliver full coverage, we need to be prepared to repeatedly run new sims as changes are made to the device up until it is frozen for tape-out. This is the subject of the next section: regression.

2.4.4. Device Regression

Curiously, the name given to re-running a full (or partial) set of simulations in order to find out if its behavior has regressed — i.e. deviated from its specification — is *regression.*[14] The subject of device regression is interesting because, with the advent of HLVLs, controversy has developed over its purpose and how to achieve its goals. First, I examine the purpose of running regressions. Then, I review the classical regression and explain how it differs from a regression performed using an autonomous verification environment.

[14] A French national I once worked for, just learning colloquial English, always referred to re-running simulations as "non-regression."

Finally, a recommended regression flow is discussed.

The dictionary definition of regress is "to return to a previous, usually worse or less developed state." Hence, the purpose of running regressions is to detect the (re-)introduction of bugs that lead the device to a "less developed state." Some bugs have the characteristic that they are easily reintroduced into the design. They need to be caught as soon as they are inserted. There are two approaches to detecting re-injected bugs: classical regression and autonomous regression.

The classical regression is performed using a test suite incrementally constructed over a period of time. The test suite is composed of directed tests specified by a test plan. Each test verifies a particular feature or function of the device. The selection criteria for adding a test to the regression test suite include that it:

- verifies fundamental behavior
- exercises a lot of the design in very few cycles
- has exposed one or more bugs in the past

Contrasted against the classical regression is the autonomous regression. An autonomous regression is performed by an autonomous verification environment, characterized by generation, checking and coverage aspects. Hundreds to thousands of copies of this environment are dispatched to a regression ranch each evening,[15] each differing from the next only in its initial random generation seed. I use the term "symmetrical simulation" to refer to a simulation whose inputs are identical to another simulation, with the exception of its random generation seed. Each regression contributes to the coverage goals of the design: functional, code and assertion. The bugs that have been found to date have been exposed by simulating until the coverage goals are achieved. The associated checkers ensure that device misbehavior does not go undetected.

Autonomous regression is preferred over classical regression because it makes use of a self-contained verification environment, dispatches symmetri-

---

[15] In Texas we have *simulation ranches*. In other parts, you might call them *simulation farms*.

[16] See chapter 7, "Coverage-Driven Verification."

cal simulations and is fully coverage-driven.[16]

2.5. Summary

In this chapter I defined functional verification, explained the difference between verification and test and outlined the functional verification process. The process is composed of writing a verification, implementing it, bringing up the DUV and running regressions.

The verification plan must define the scope of the verification problem — using functional, code and assertion coverage — and specify how the problem will be solved in the stimulus generation and response checking aspects.

The verification environment is implemented using pre-built verification components wherever possible. When new components are required, they should be implemented as reusable verification IP, according to ***e***RM guidelines when ***e*** is the implementation language.

Device bring-up is used to get the device simulating and expose any gross errors. An iterative cycle of "make an assumption, validate the assumption" is followed, incrementally increasing the complexity of the applied stimuli until the device is ready for full regression.

The purpose of regression is to detect the reintroduction of bugs into the design. Regression methodology has progressed from running a directed regression test suite to dispatching symmetrical autonomous simulations. The autonomous simulations are coverage-driven by the functional, code and assertion coverage goals of the verification plan.

3. Measuring Verification Coverage

In order to measure verification progress, we measure verification coverage because verification coverage defines the extent of the verification problem. In order to measure anything however, we need metrics. In this chapter I define coverage metrics and a useful taxonomy for their classification. Using this taxonomy, I introduce the notion of a coverage space and define four orthogonal spaces into which various kinds of coverage may be classified.

3.1. Coverage Metrics

A coverage metric is a parameter of the verification environment or device useful for assessing verification progress in some dimension. We may classify a coverage metric according its kind and its source. By kind, I mean whether or not a metric is implied from the verification interface[1] or is explicitly defined. Hence, a metric kind is either implicit or explicit.

The second classification is the source of a metric, which has a strong bearing on what verification progress we may infer from its value. I will consider two sources, each at a different abstraction level: implementation and specification. An implementation metric is one taken from the device implementation, for example the RTL. A specification metric is a metric extracted from one of the device specifications: the functional or design specification.

These two classifications of coverage metrics define four metrics, as illustrated in table 3-1 below.

---

[1] See chapter 1, "The Language of Coverage," for the definition of "verification interface."

Metric Source		Implicit	Explicit
	Specification	Implicit Specification	Explicit Specification
	Implementation	Implicit Implementation	Explicit Implementation

Metric Kind

Table 3-1 Coverage Metric Taxonomy

The following four sections explore in greater detail these coverage metric classifications. The first two describe metric kinds while the second two describe metric sources.

3.1.1. Implicit Metrics

An implicit coverage metric is inherent in the representation of the abstraction level from which the metric is taken. For example, at the RTL abstraction level, hardware description language structures may be implicit metrics. A Verilog statement is an implicit metric because statements are a base element of the Verilog language. The subexpressions of a VHDL Boolean expression in an "if" statement may be implicit metrics. The same metrics may be applied to the verification environment, independent of its implementation language.

Another abstraction level from which implicit metrics may derived is the device specification. The implicit metrics of a natural language specification include chapter, paragraph, sentence, table and figure.

3.1.2. Explicit Metrics

An explicit coverage metric is invented, defined or selected by the engineer. It is usually selected from a natural language specification but it could also be chosen from an implementation. Explicit metrics are typically used as components for modeling device behavior.

Examples of explicit metrics from the CPU application domain are:

- opcode
- register
- address
- addressing mode
- execution mode

Examples from an ethernet application are:

- preamble
- start frame
- source address
- destination address
- length
- CRC

The next two sections describe the two sources of coverage metrics: the device specifications and its implementation.

3.1.3. Specification Metrics

A specification coverage metric is a metric derived from one of the device specifications: the functional specification or the design specification. Since these specifications record the intended behavior of the device, the metrics extracted from them are parameters or attributes of that behavior. In effect, these metrics quantify the device behavior, translating its description from a somewhat ambiguous natural language abstraction to a precise specification.

Some examples of specification metrics are:

- instruction opcode
- packet header
- processing latency
- processing throughput

3.1.4. Implementation Metrics

An implementation coverage metric is a metric derived from the implementation of the device. The device implementation is distinguished from its specification in that the implementation is much less abstract, less ambiguous and machine readable. Recall that the design specification describes the microarchitecture of the device. However, it still leaves many design choices to the designer writing the RTL. Metrics derived from the implementation are design choices which are not present in its specification because they are an implementation choice.

These are a few examples of implementation metrics:

- one-hot mux select value
- finite state machine state
- pipeline latency
- pipeline throughput
- bandwidth

Having introduced the four types of coverage metrics, let's turn our attention to the coverage spaces which are defined from these metrics.

3.2. Coverage Spaces

What do we mean by a "coverage space?" As defined in chapter 1, "The Language of Coverage," a coverage space is a multi-dimensional region defined by its attributes and their values. A coverage space is often referred to as a coverage model because it captures the behavior of the device at some abstraction level.[2] The kind and source of coverage metric, each having two

[2] Simone Santini eloquently defined an abstraction in the May 2003 issue of "Computer" magazine: "An abstraction ignores the details that distinguish specific instances and considers only those that unify them as a class."

values, define four types of coverage spaces: implicit implementation, implicit specification, explicit implementation and explicit specification. Each of these coverage spaces is discussed in the following sections.

3.2.1. Implicit Implementation Coverage Space

The first type of coverage space I will discuss is the implicit implementation coverage space. An implicit implementation coverage space is composed of metrics inherent in what we are measuring and extracted from the implementation of the device. The implementation of the device we are concerned with is its RTL abstraction, commonly written in Verilog, SystemVerilog, VHDL or a proprietary hardware description language. Each of these languages is defined by a grammar which specifies the structure of the language elements and associates semantic meaning to those constructs.

The RTL implementation of the device may be considered to be mapped onto the structure of its implementation language. As such, if we record the language constructs exercised as we simulate the device, we gain insight into how well the device has been verified.[3]

Code coverage and structural coverage define implicit implementation coverage spaces.[4]

Because code and structural coverage is measured at a low abstraction level, more detail than is often needed is reported. This leads to the need for filtering of reported results in order to gain insight into how well the device has been exercised.

3.2.2. Implicit Specification Coverage Space

An implicit specification coverage space is composed of metrics inherent in the abstraction measured and extracted from one of the device specifications. The abstraction measured is a natural language, typically English. The device specifications include the functional specification and the design specification.

In order for a coverage mechanism to be classified as defining an implicit specification coverage space, the metrics would have to be inferred

[3] I assume here that an orthogonal mechanism is in place to detect device misbehavior such as data and temporal checkers.

[4] Code coverage is the subject of chapter 5.

from one or more of the device specifications. The mechanism would have to parse the natural language, recognize grammatical language structure and context and extract metrics and their relationships. Metrics could be drawn from both the syntactic elements of the language or from its semantic meaning (quite challenging!). Potential syntactic metrics include:

- chapter
- section
- paragraph
- sentence
- sentence flow
- word
- footnote

while semantic metrics might be:

- corner case
- boundary condition
- special-purpose
- finite state machine
- arbitrator or arbiter
- queue

As of the date this book was written, I am aware of no implementations of a program which derives implicit specification coverage spaces.

3.2.3. Explicit Implementation Coverage Space

An explicit implementation coverage space is composed of metrics invented by the engineer and derived from the device implementation. These metrics should reflect design choices made by the designer, unspecified in the design specification, which cannot be inferred from RTL using code coverage.

For example, the device may have a pipelined bus interface that supports three outstanding transactions. The coverage space for this interface must include one-deep (un-pipelined), two-deep and three-deep transactions. If the bus interface is not specified at this level of detail in the design specification, this space is an explicit implementation coverage space because the

RTL must be examined to ascertain the pipeline parameters.

Another example is a tagged bus interface in which transaction responses may return out of order (return order differs from the request order). The coverage space needs to include permutations of in-order and out-of-order responses.

A third example is a functional coverage model of the device microarchitecture (sometimes referred to as the design architecture). This also is a explicit implementation coverage space. In the next chapter, "Functional Coverage," you will learn how to design a functional coverage model.

Since a functional coverage model captures the precise level of detail required to determine verification progress, no filtering of reported results is required. If the reported results are found to be more voluminous than necessary, the associated coverage model may be pruned. This is true of both explicit coverage spaces: implementation and specification.

3.2.4. Explicit Specification Coverage Space

An explicit specification coverage space is composed of metrics invented by the engineer and derived from one of the device specifications. By "invented," I mean each metric is chosen by the DV engineer from a large population of potential metrics discussed in the device specifications. The metrics are selected from both the data domain and the time domain. Data domain metrics are values or ranges representing information processed by the device. Time domain metrics are values or ranges of parameters of sequential device behavior. As we'll see in the next chapter, each metric represents an orthogonal parameter of the device behavior to be related to others in one or more models.

A functional coverage model of the device defines an explicit specification coverage space. The coverage model may be an input, output, input/output or internal coverage model. The next chapter, "Functional Coverage," discusses in detail the design and implementation of functional coverage models.

The four coverage spaces fit into the coverage metric taxonomy as illustrated in table 3-2 below.

		Metric Kind	
		Implicit	Explicit
Metric Source	Specification	Implicit specification coverage	Specification functional coverage
	Implementation	Code and structural coverage	Implementation functional coverage

Table 3-2 Coverage Spaces

Each of the coverage spaces is used to observe device behavior from a different perspective. Specification functional coverage indicates what features and capabilities of the DUV, as documented in its specification, have been exercised on its input, output and internal interfaces. Implementation functional coverage reports scenarios observed at the register transfer level. Code and structural coverage offer insight into how extensively the implementation of the device has been exercised. Implicit specification coverage would tell us how much of the device specification has been exercised. Unfortunately, this has not yet been implemented to the best of my knowledge.

3.3. Summary

In this chapter I explained why we use verification coverage as a measure of verification progress. I introduced the concept of coverage metrics and classified them into implicit, explicit, implementation and specification metrics. This classification was used to build a taxonomy of coverage spaces, regions of behavior defined by metrics. Lastly, each of the kinds of coverage were placed in this taxonomy.

4. Functional Coverage

In the previous chapter, "Measuring Verification Coverage," you learned how to classify coverage and that functional coverage defines an explicit implementation or specification coverage space, depending upon the source of the coverage metrics. If the metrics are derived from the implementation itself, it defines an explicit implementation coverage space. If the metrics are derived from one of the device specifications, the functional coverage defines an explicit specification coverage space.

In this chapter, I explore the use of functional coverage to model device behavior at various verification interfaces. You will learn the process of top-level design, detailed design and implementation of functional coverage models. The specific requirements of temporal coverage measurement and finite state machine coverage will also be addressed.

4.1. Coverage Modeling

The purpose of measuring functional coverage is to measure verification progress from the perspective of the functional requirements of the device. The functional requirements are imposed on both the inputs and outputs of the device — and their interrelationships — by the device specifications.[1] The input requirements dictate the full data and temporal scope of input stimuli to be processed by the device. The output requirements specify the complete set of data and temporal responses to be observed. The input/output requirements specify all stimulus/response permutations which must be observed to meet black-box device requirements.

Since the full behavior of a device may be defined by these input, output and input/output requirements, a functional coverage space which cap-

[1] Functional specification and design (microarchitecture) specification.

tures these requirements is referred to as a coverage model.[2] The degree to which a coverage model captures these requirements is defined to be its fidelity.

The fidelity of a model determines how closely the model defines the actual behavioral requirements of the device. This is the abstraction gap between the coverage model and the device. If we are modeling a control register having 18 specified values and the coverage model defines all 18 values, their is no abstraction gap so this is a high fidelity model. However, if a 32-bit address bus defines all 2^{32} values and an associated coverage model groups those values into 16 ranges, a substantial abstraction gap is introduced. Hence, this would be a lower fidelity model.

In addition to the abstraction gap between the coverage model and the device, another source of fidelity loss is omitting functional relationships from the model. If two bits, such as CR_3 and CR_7 of the control register mentioned above, have a dependency such as "CR_3 may only be one if CR_7 is one," yet this dependency is not reflected in an associated coverage model, that model is of a lower fidelity than a model that captures this relationship.

Before diving into the details of designing a real coverage model, I first illustrate the whole process from beginning to end with a simple example.

4.2. Coverage Model Example

The following is a brief functional specification of the device. The "device" to be modeled is a wood burning stove we use to heat our house in the winter. It has three controls: logs, thermostat and damper. The logs fuel the stove, which can only burn three to six logs at a time. The thermostat modulates the air intake to maintain the desired stove temperature, ranging from 200° to 800° F in 100° increments. The damper either directs the combustion exhaust straight out the stove pipe or through a catalytic converter. It may be either open or closed.

The rules for operating the stove are:

1. The damper must not be closed unless the stove is at least 400°.
2. The damper must be closed once the stove reaches 700°.

[2] Management of the internal state of the device is implied by these I/O requirements and may be observed using code coverage or an internal functional coverage model.

3. No more than four logs should be used for a 200° to 400° stove.
4. Five logs are required for 500°.
5. Six logs are required for the stove to reach 700° or hotter.

The semantic description of the wood stove coverage model is: *Record all valid operating conditions of the wood stove defined by logs, thermostat and damper.* The parameters, known as attributes, to be used to design the model are listed below (table 4-1):

Attribute	**Values**
logs	3 - 6
thermostat	200° - 800°
damper	closed, open

Table 4-1 Wood Stove Attributes

The operating conditions for the stove, defined by its rules, are captured in the following table (4-2):

Attribute		Logs	Thermostat	Damper
Value		3-6	200° - 800°	closed, open
Sampling Time		wood loaded	thermostat set	damper changed
Correlation Time	15 min	3, 4	200°, 300°	open
		3, 4	400°	*
		5	500°	*
		6	600°	*
		6	700°, 800°	closed

Table 4-2 Wood Stove Coverage Model

The attributes of the model are listed in the first row. All of their possible values are listed in the second row. The time each value will be recorded, or sampled, is listed in the third row. The remaining rows define the relationships among the attributes (model structure) and when the groups of attributes should be recorded (correlation time).

For example, we must observe three and four logs used to operate the stove at both 200° and 300°, with the damper open (first correlation time row). We must also observe three and four logs used to operate the stove at 400°, with the damper open and with the damper closed (second correlation time row), and so on. The asterisk means all values of the attribute in this column.

The total number of coverage points in this model is the sum of the points defined in each of the correlation time rows. The number of points defined in a correlation time row is the product of the number of attribute values specified in each of its columns, as illustrated below:

Logs	Thermostat	Damper	Number of Points
3, 4	200°, 300°	open	4
3, 4	400°	closed, open	4
5	500°	closed, open	2
6	600°	closed, open	2
6	700°, 800°	closed	+ 2
			16

Hence, this model defines a coverage space of 16 points.

Figure 4-1 below illustrates the structure of the wood stove coverage model. Each attribute occupies a level in the tree. Attribute values label each arc.

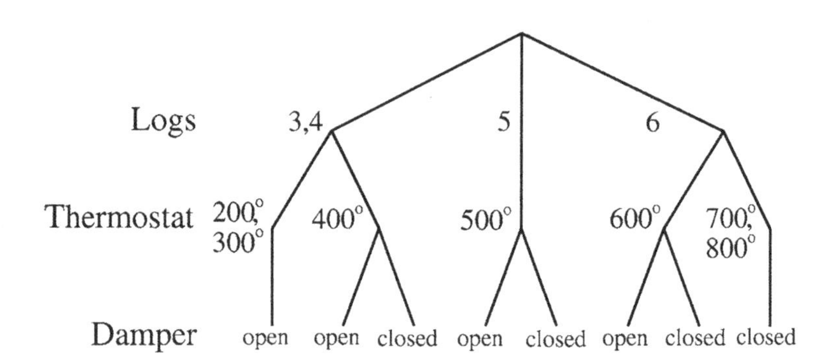

Figure 4-1 Wood Stove Coverage Model

The coverage model is implemented in *e* as follows. The first line extends an agent, `ap_stove_agent_u`, that monitors the wood stove.

Lines 2 to 7 define a coverage group of three simple items — `logs`, `stat` (thermostat) and `damper` — and one cross item — `logs × stat × damper`. Each item refers to a field declared below. Line 9 defines the correlation event for the coverage group, `wood_stove_e`. It is emitted every 15 minutes. Lines 11 to 13 declare the attributes to be sampled: `logs`, `stat` and `damper`.

```
 1  extend ap_stove_agent_u {
 2    cover wood_stove_e is {
 3      item logs;
 4      item stat;
 5      item damper;
 6      cross logs, stat, damper
 7    };
 8
 9    event wood_stove_e;
10
11    logs   : uint [3..6];
12    stat   : uint [200..800];
13    damper : [OPEN, CLOSED];
```

Lines 15 to 26 restrict the permutations of the attributes to those defined by the model. The restriction is implemented as the negation of the valid attribute relationships: `using ignore = not (`*valid conditions*`)`.

```
15    cover wood_stove_e is also {
16      item cross__logs__stat__damper
17      using also ignore = not (
18        (logs in [3..4] and stat in [200..300]
19          and damper == OPEN) or
20        (logs in [3..4] and stat == 400) or
21        (logs == 5      and stat == 500) or
22        (logs == 6      and stat == 600) or
23        (logs == 6      and stat in [700..800]
24          and damper == CLOSED)
25      )
26    };
```

Finally, the sampling events are defined for `logs`, `stat` and `damper`. Whenever event `wood_loaded_e` is emitted, the number of logs in the stove is captured from the `logcnt` signal.

```
28     event wood_loaded_e;

30     on wood_loaded_e {
31       logs = 'top/chamber/logcnt'
32     };
```

Whenever thermostat_set_e is emitted, the thermostat temperature setting is copied from top/statpos.

```
33     event thermostat_set_e;

35     on thermostat_set_e {
36       stat = 'top/statpos'
37     };
```

And, whenever damper_changed_e is emitted, the damper position is captured from top/dpos.

```
38     event damper_changed_e;

40     on damper_changed_e {
41       var d := 'top/dpos';
42       case d {
43         0: {damper = OPEN};
44         1: {damper = CLOSED}
45       }
46     };
47   } //extend ap_stove_event_u//
```

Having de-mystified the overall process of designing and implementing a functional coverage model, in the following sections I explain in detail the steps required to design and implement a coverage model using a real world example.

4.3. Top-Level Design

The design process is divided into two phases: top-level design and detailed design. Top-level design concerns itself with describing the semantics of the model, identifying the attributes and specifying the relationships among these attributes which characterize the device. Detailed design maps these attributes and their relationships into the verification environment.

The semantics of the model is an English description of what is modeled, sometimes called a story. An example for an input coverage model is: *The instruction decoder must decode every opcode, in every addressing mode with all permutations of operand registers.* An output model example is: *We must observe the packet processor write a sequence of packets of the same priority, where the sequence length varies from one to 255.*

Once the semantic description is written, the second step in designing a coverage model is identifying attributes. But first, what is meant by an *attribute?* Referencing chapter one, "The Language of Coverage," an attribute is defined as *a parameter or dimension of the model.* In other words, an attribute identified from one of the device specifications is a parameter of the device such as configuration mode, instruction opcode, control field value or packet length. As we'll see later in this chapter, an attribute in a coverage model describes part of its structure. The attribute may define a dimension of a matrix model or a level in a hierarchical coverage model. The attributes we are initially concerned with are those extracted from the device specifications.

4.3.1. Attribute Identification

The second step in designing a coverage model is identifying attributes, their values and the times they should be sampled. The most effective way to identify attributes is in a brainstorming session among several verification engineers and designers familiar with the device specifications. If you need to design a coverage model single-handedly, make sure you are familiar with the specifications before embarking on the endeavor. You'll need to visit each section of device requirements and record parameters that may be used to quantify the description of those features. The specifications I reference in the remainder of this chapter are those that define the intended behavior of processors implementing the ubiquitous IA-32 instruction set architecture. They are the "IA-32 Intel Architecture Software Developer's Manual, Volume 1: Basic Architecture"[3] and the "IA-32 Intel Architecture Software Developer's Manual, Volume 2: Instruction Set Reference."[4]

The IA-32 instruction set architecture, more commonly referred to as the x86 architecture, has a rich feature set from which attributes may be

[3] In November 2003, this manual was available online at http://developer.-intel.com/design/pentium4/manuals/24547012.pdf.

[4] http://developer.intel.com/design/pentium4/manuals/24547112.pdf.

drawn. They include:

- execution mode
- instruction opcode
- general purpose register
- instruction pointer (EIP)
- flags register (EFLAGS)

Execution mode defines what subset of the architecture is available to the running program. *Instruction opcode* is the primary field of an instruction which encodes its operation. The *general purpose registers* (GPR, which are really not so general) are traditional, fast-access storage elements used for instruction operands. The *instruction pointer register* (EIP) specifies the address offset of the next instruction to be executed. The *flags register* (EFLAGS) contains status flags, a control flag and system flags. After the attributes are selected, the next step is to choose the attribute values to be recorded.

An attribute value may be a single value or a range of values. The criteria for selecting attribute values depend upon the kind of attribute. If the attribute is a control field, each of the control values should be enumerated. If the attribute is a data field, the values to be selected depend upon the nature of the data and how it is processed by the device. For example, if a data field is the source operand of a move instruction, the behavior of the instruction is likely not dependent on the operand value. On the other hand, if it is one of the source operands of an arithmetic instruction, the value of the operand strongly influences the instruction behavior. Let's use the attributes chosen above to illustrate attribute value selection.

The attribute *execution mode* is a control field so we specify each of its possible values: real address mode, protected mode and system management mode. Although "virtual-8086 mode" is sometimes considered an execution mode, it is actually a protected mode attribute, so it is excluded it from the execution mode values.

The attribute *instruction opcode* is a one- or two-byte encoding of an instruction operation within the instruction. In addition, part of the opcode may be encoded in three bits of another instruction field called the ModR/M byte. $\text{ModR/M}_{5:3}$ is known as the reg/opcode field of ModR/M.[5] Here are

[5] Throughout this book I use little endian bit notation where bit zero is the least significant bit of a number.

several instruction opcode examples from appendix A of the instruction set reference manual:

Instruction Name	Opcode	Notes
ADD	00 - 05	1-byte opcode
MOV	A0 - A5, B0 - B7, C6, C7, 88 - 8C, 8E, B8 - BF	1-byte opcode
LAR	0F02	2-byte opcode
RDMSR	0F32	2-byte opcode
ADC	80 /2	1-byte + $ModR/M_{5:3}$
SBB	80 /3	1-byte + $ModR/M_{5:3}$

The attribute values are 00, 01, 02, ... 0F02, 0F32, ... 80 /2, 80 /3, etc.

The attribute *general purpose register* specifies one of eight 32-bit registers, eight 16-bit registers or eight 8-bit registers:

32-bit Register	16-bit Register	8-bit Register	8-bit Register
EAX	AX	AH	AL
EBX	BX	BH	BL
ECX	CX	CH	CL
EDX	DX	DH	DL
ESI	SI		
EDI	DI		
EBP	BP		
ESP	SP		

Each 16-bit register is an alias for the lower 16 bits of its corresponding 32-bit register. For example, the AX register is aliased to $EAX_{15:0}$ and BX is aliased to $EBX_{15:0}$. In total, there are 24 general purpose registers, hence 24 GPR attribute values.

The attribute *instruction pointer register* (EIP) is a 32-bit register containing an unsigned byte offset in the code segment to the next instruction to be executed. The only significant boundary condition values of this register are zero, one, $2^{32} - 1$ and 2^{32}. For completeness, we also sample values throughout the range 2 to $2^{32} - 2$. These are the attribute values.

The attribute *flags register* (EFLAGS) is composed of a set of status flags, one control flag, a set of system flags and a set of reserved bits. The following table documents the EFLAGS register:

EFLAGS Position	Flag Name	Flag Type	Description
0	CF	status	carry flag
1	----	-----	(reserved)
2	PF	status	parity flag
3	----	-----	(reserved)
4	AF	status	auxiliary carry flag
5	----	-----	(reserved)
6	ZF	status	zero flag
7	SF	status	sign flag
8	TF	system	trap flag
9	IF	system	interrupt enable flag
10	DF	control	direction flag
11	OF	system	overflow flag
13:12	IOPL	system	I/O privilege level
14	NT	system	nested task
15	----	-----	(reserved)
16	RF	system	resume flag
17	VM	system	virtual-8086 mode
18	AC	system	alignment check
19	VIF	system	virtual interrupt flag
20	VIP	system	virtual interrupt pending
21	ID	system	ID flag
31:22	----	-----	(reserved)

Each of these attributes is of a control nature so we define attribute values for each of their possible values. The reserved bits of EFLAGS are excluded from the values recorded because their values and behavior are not defined.

The last concern we have with attribute identification is selecting the sampling time for each attribute. An attribute may be sampled at a variety of times: each time its value is changed, on every cycle or in conjunction with another event. How do we decide when and how often it should be sampled?

There are several things to consider when choosing an attribute sampling time. First of all, a value should be sampled no more frequently than it is updated. A system control register bit, for example, may only be written when an associated control register write instruction is executed. If the monitor sampling the register has access to the execute stage of the processor, it could read the control register bit only when the write instruction is executed. It may also have direct access to the control register write logic in which case it could simply watch the register write enable line.

An attribute should also be sampled no more frequently than the frequency at which it interacts with other attributes. This "interaction frequency" is the inverse of the attribute correlation time. Because attributes interact with others, they create attribute relationships. Attribute correlation times and attribute relationships are discussed in the next section. A data or control attribute is usually sampled whenever it is updated. For example, a register attribute would be sampled when the register is written. A control bit attribute would be sampled whenever the value of the control bit is altered.

With these considerations in mind, let's select the sampling times for the IA-32 attributes.

The execution mode of an IA-32 processor is determined by the value of the protected mode enable bit of control register zero (CR0.PE), the SMI# pin and the resume (RSM) instruction. CR0.PE should be sampled whenever the CR0 register is written or combined with another attribute. The SMI# pin should be sampled whenever it is asserted. The RSM instruction should be sampled each time it is executed. Figure 4-2 below illustrates how PE, SMI# and RSM cause changes to the execution mode.

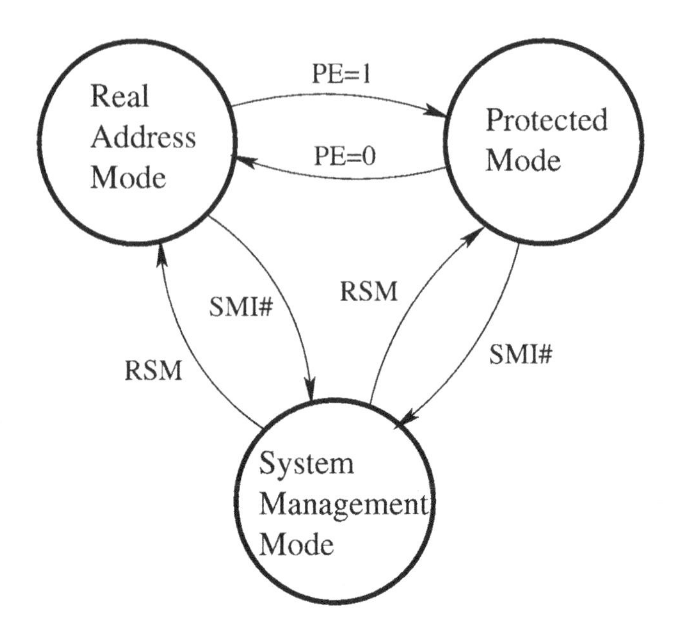

Figure 4-2 IA-32 Execution Mode FSM

An instruction opcode should be sampled whenever an instruction is completed (i.e. retired). The opcode is not sampled at the time an instruction is fetched because some instructions are speculatively fetched, but not executed. Any general purpose register, the instruction pointer (EIP) or the flags register (EFLAGS) should be sampled whenever it is written. None of these attributes should be sampled more frequently than the least frequently sampled attribute with which it is related (as explained in "Attribute Relationships" below.

Now that the attributes, values and sampling times needed to construct the coverage model have been defined, the next step is identifying the relationships among the attributes.

4.3.2. Attribute Relationships

Attribute relationships should reflect attribute interactions, dependencies and device behavior specified by combinations of attributes. Each of these behavioral relationships are reflected in device logic purposefully designed to meet these requirements or in particular scenarios described in the device specifications. Associated with each attribute relationship is a correlation time. The correlation time is a moment two or more attributes participate in decision logic of the device. For example, when an IA-32 processor needs to know if it is executing in virtual-8086 mode, CR0.PE and EFLAGS.VM are sampled.[5.1] Therefore, CR0.PE and EFLAGS.VM are correlated when we need to know if we are in virtual-8086 mode.

The structure of the relationship may be of one of three types: matrix, hierarchical or hybrid. A matrix model considers each attribute to be a dimension of a matrix, where the number of dimensions is defined by the number of attributes. The values along each axis are the values of the corresponding attribute. Figure 4-3 below illustrates a two dimensional matrix model.

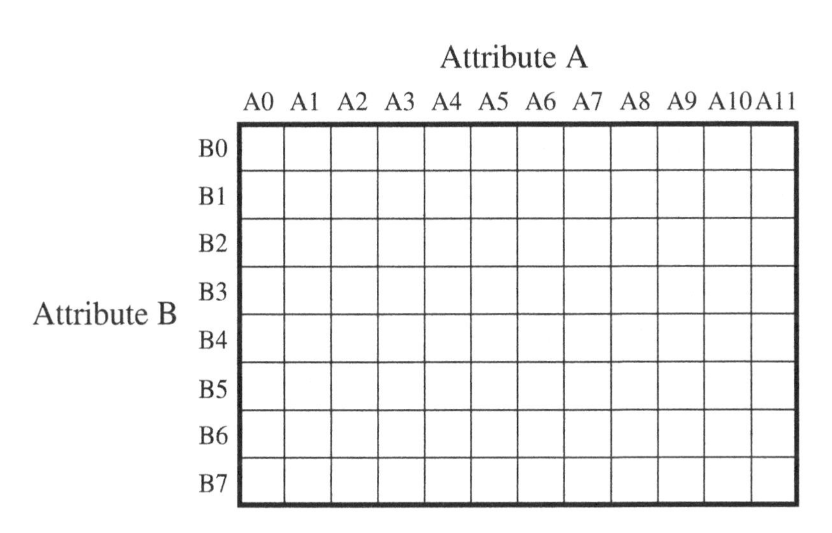

Figure 4-3 Matrix Coverage Model

The two attributes are labeled "Attribute A" and "Attribute B." Attribute A has twelve values, A0 through A11 while attribute B has eight values, B0 through B7. An attribute pair, (A_n, B_m) defines a point within this two dimensional coverage space. This matrix model defines 96 points.

A hierarchical model is structured like an inverted tree with its root at the top.[6] It is a directed graph whose nodes are attribute values and whose

[5.1] This is a simplification for the sake of discussion. In reality, all IA-32 processor implementations maintain a virtual-8086 state bit.

[6] Why do engineers and computer scientists always draw trees upside down?

edges indicate a relationship between one attribute value and another. One attribute is represented at each level in the tree, with the primary controlling attribute at the root and each successive attribute one level lower. In figure 4-4 below, I illustrate a hierarchical coverage model defined by three attributes: "Attribute A," "Attribute B" and "Attribute C."

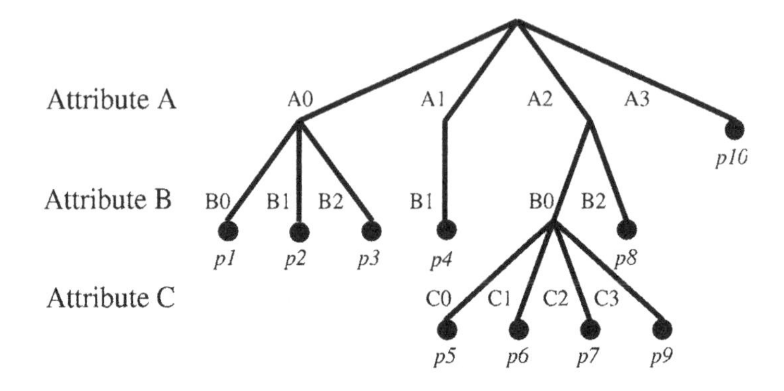

Figure 4-4 Hierarchical Coverage Model

Attribute A is the primary controlling attribute and has four values, A0 through A3. Attribute B is a secondary attribute and has three values, B0, B1 and B2. Attribute C is a tertiary attribute having four values, C0 through C3. In this model, an attribute 3-tuple $(A_{n,} B_{m,} C_k)$ defines one of the ten points p_i in the coverage space. For example, p_7 is defined by (A2, B0, C2), the path to this point.

A hybrid model is composed of a blend of matrix and hierarchical regions, where fully permuted attribute combinations are structured as a matrix and irregular attribute relationships are structured hierarchically. The matrix subregions may be leaf nodes of a base hierarchical model or hierarchical subregions may occupy nodes within a base matrix model. Figure 4-5 below illustrates a hybrid model of the first type, matrix subregions are leaf nodes of a base hierarchical model.

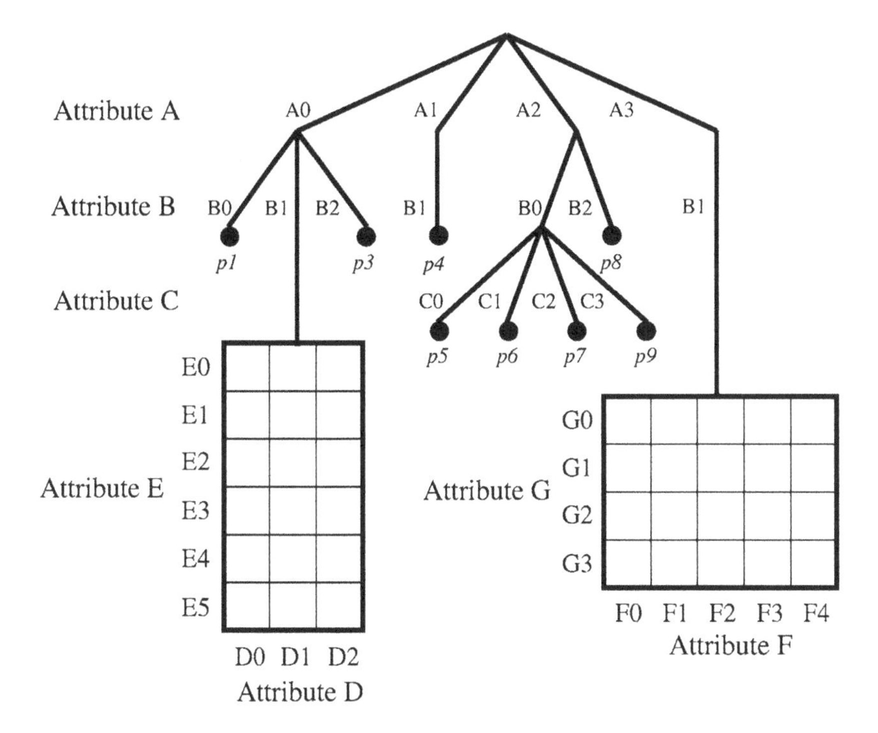

Figure 4-5 Hybrid Coverage Model

Attributes A, B and C are defined the same as the previous hierarchical model, having four, three and four values respectively. Attribute D has three values — D0, D1, and D2 — and attribute E has six values — E0 through E5. These two attributes define a three by six matrix subregion of the hierarchical model. Attributes F and G likewise define a five by four matrix subregion.

This model defines a total of 46 points: 8 leaf nodes (p_1, p_3, p_4, p_5, p_6, p_7, p_8, p_9), 18 points in the attribute D/E matrix and 20 points in the attribute F/G matrix.

Although the structure of the coverage model should reflect the underlying relationships among the attributes from which it is composed, there are times when the relationships may be modeled using more than one of these model structures. What are the pros and cons of each structure in terms of model fidelity and implementation effort?

The interrelationship between model fidelity and model structure is determined by how closely the inherent relationships among the attributes of a model reflect a particular model structure. For example, if the semantic description of a coverage model specifies that all register pairs of two-operand instructions are to be recorded, the full permutation of registers is a matrix. On the other hand, if the description states that the first operand register must be observed with second operand registers sequentially one, two and three greater than the first, as illustrated in table 4-3 below:

First Operand Register	Second Operand Register
0	1, 2, 3
1	2, 3, 4
2	3, 4, 5
3	4, 5, 6
4	5, 6, 7
5	6, 7, 0
6	7, 0, 1
7	0, 1, 2

Table 4-3 Register Operand Pairs Specification

it reflects the following hierarchical structure (figure 4-6):

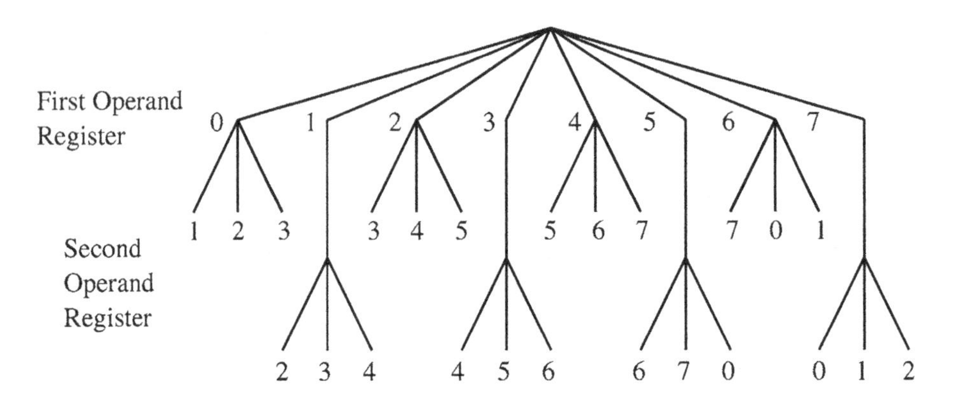

Figure 4-6 Register Operand Pairs Structure

The matrix model requires the least effort to design and implement because of its symmetry. Each of the interacting attributes is specified, along with its corresponding values. Because the matrix model is so easy to build, engineers often choose it as the path of least resistance even though another model is more applicable. If the full size of a model, structured as a matrix, is relatively small (1,000 points or less) and the extraneous coverage points are not invalid or impossible to reach, the time saved over designing a more precise hierarchical or hybrid model is worth the loss of fidelity. However, if the matrix model would be quite large (100,000 or more points) or it would define many invalid points, a hierarchical or hybrid model should be chosen instead.

The hierarchical model requires more effort to design because, not only must the attributes and their values be specified, specific relationships among attributes must also be defined. These relationships are often quite irregular and complex, requiring many lines to enumerate. Nonetheless, the complexity reflected in the hierarchical model is inherent in the attribute relationships described by the device specification or reflected in the RTL implementation. Although it may be simplified using a lower fidelity model, it cannot be avoided. The strength of the hierarchical model is that it can precisely define attribute relationships, dramatically reducing the size of a coverage model and providing deeper insight into observed device behaviors.

The hybrid model requires a design effort comparable to the hierarchical model. Again, attributes, values and individual relationships must be enumerated. However, some regions of a hybrid model are quite regular and are represented by a matrix structure. The hybrid model usually reflects the most precise input, output or internal device behavior because of the nature of designs. They are a blend of symmetric values with a few exceptions tossed in.

Although implementing these models is discussed later in this chapter, I should emphasize I am discussing coverage model structure here, not its implementation structure. A hierarchical or hybrid model may be implemented using an ***e*** cross coverage group, so long as the requisite `using ignore` and `using illegal` options are used. Even through the name "cross coverage group" strongly implies a matrix structure, it is often used to implement models of all three structures.

Having examined coverage model structures in some depth, let's turn to an example of a behavioral relationship. Among the attributes we've selected from the IA-32 architecture, execution mode interacts with the EFLAGS VM bit because virtual-8086 mode may only be entered from protected mode. We will capture the attributes, their values, sampling time, correlation time and relationships in a table (4-4):

<table>
<tr><td colspan="2">Attribute</td><td>Execution Mode</td><td>VM</td></tr>
<tr><td colspan="2">Value</td><td>real address mode, protected mode, system management mode</td><td>0, 1</td></tr>
<tr><td colspan="2">Sampling Time</td><td>CR0 write</td><td>EFLAGS.VM write</td></tr>
<tr><td rowspan="2">Corre-lation Time</td><td rowspan="2">CR0 or EFLAGS VM write</td><td>protected mode</td><td>0</td></tr>
<tr><td>protected mode</td><td>1</td></tr>
</table>

Table 4-4 Virtual-8086 Mode Coverage Model

It is composed of four primary rows and three columns.[7] The primary rows are attribute, attribute value, attribute sampling time and attribute correlation times and relationships. The columns are row title, execution mode values and VM values. In the attribute row, the name of each attribute that contributes to the coverage model is listed. In the value row, all of the selected values for each attribute are recorded. In the sampling time row, the time at which the attribute value should be recorded is listed. In the correlation time row, the first column to the right of "Correlation Time" lists the correlation time for the attributes related on that row. The remaining columns to the right record relationships among the attributes.

This table specifies a coverage model composed of two attributes: execution mode and VM. The selected values of execution mode are "real address mode," "protected mode" and "system management mode." The values of VM are 0 and 1. The sampling time of execution mode is a CR0 write. The sampling time of VM is a write to VM. The two attributes are always correlated at the same time: whenever CR0 or EFLAGS.VM is written.[8] The

[7] The rows and columns of the table could be exchanged to make room for more attributes.

[8] It is possible to define a different correlation time for each attribute relationship row.

two attribute relationships defined by the model define only two points: {protected mode, 0} and {protected mode, 1}.

This table describes a coverage model composed of only two coverage points: {execution mode, VM} = {protected, 0} and {protected, 1}. When each of the values of an attribute contribute to the model as above, we use an asterisk ("*") to represent all values:

Attribute		Execution Mode	VM
Value		real address mode, protected mode, system management mode	0, 1
Sampling Time		CR0 write	EFLAGS.VM write
Correlation Time	CR0 or EFLAGS VM write	protected mode	*

Table 4-5 Abbreviated Virtual-8086 Mode Coverage Model

Returning to specific IA-32 attribute relationships, another model of interest is one which records the use of a general purpose register as a destination register for each permutation of the arithmetic flags. (This is its semantic description.) Although the IA-32 specifications do not explicitly state that these interact, a discussion with one of the design engineers revealed that the arithmetic flag logic in our processor is optimized for particular registers. This model may be described as:

Attribute		Destination GPR	CF	PF	AF	ZF	SF	OF
Value		EAX, EBX ECX, EDX ESI, EDI EBP, ESP AX, BX CX, DX SI, DI BP, SP AH, BH CH, DH AL, BL CL, DL	0,1	0,1	0,1	0,1	0,1	0,1
Sampling Time		t_{gpr}	t_{CF}	t_{PF}	t_{AF}	t_{ZF}	t_{SF}	t_{OF}
Corre-lation Time	t_{ew}	*	*	*	*	*	*	*

t_{gpr} – register is written t_{ew} – EFLAGS is written
t_{CF} – CF is written t_{PF} – PF is written
t_{AF} – AF is written t_{ZF} – ZF is written
t_{SF} – SF is written t_{OF} – OF is written

Table 4-6 GPR/Arithmetic Flags Coverage Model

It has seven attributes (general purpose [destination] register, CF, PF, AF, ZF, SF, and OF), each sampled whenever its value is changed. The correlation time row indicates that the attributes are correlated whenever EFLAGS is written and that all permutations of all of the attribute values define the model. Therefore, the model has 1,536 coverage points: (24 GPRs) × (2 CF values) × (2 PF values) × (2 AF values) × (2 ZF values) × (2 SF values) × (2 OF values). Since the full permutation of attribute values defines a matrix, we use a matrix model.

To illustrate a hierarchical model, we will model a subset of the relationship between instruction opcodes and the arithmetic flags. These relationships are documented in table A-1 of the basic architecture manual, a subset of which is reproduced in table 4-7 below:

Instruction	OF	SF	ZF	AF	PF	CF
AAA	–	–	–	TM	–	M
AAD	–	M	M	–	M	–
AAM	–	M	M	–	M	–
AAS	–	–	–	TM	–	M
ADC	M	M	M	M	M	TM
ADD	M	M	M	M	M	M
AND	0	M	M	–	M	0
ARPL			M			
BOUND						
BSF/BSR	–	–	M	–	–	–
BSWAP						
BT/BTS/BTA/BTC	–	–	–	–	–	M

Table 4-7 EFLAGS Cross-Reference

where the cell entries in the flag columns have the following meanings:

T	instruction tests flag.
M	instruction modifies flag.
0	instruction resets flag.
–	instruction effect on flag is undefined.
blank	instruction does not affect flag.

Table 4-8 Instruction EFLAGS Effects

Our model will reflect all but undefined behavior from this specification. Since some instructions influence (or are influenced by) flags and others do not, a hierarchical model must be used. The model is represented by the following table (4-9):

Attribute		Instruct	OF	SF	ZF	AF	PF	CF
Value		AAA, AAD, AAM, AAS, ADC, ADD, AND, ARPL, BOUND, BSF, BSR, BSWAP, BT BTS BTA, BTC	0 1	0 1	0 1	0 1	0 1	0 1
Sampling Time		t_{inst}	t_{OF}	t_{SF}	t_{ZF}	t_{AF}	t_{PF}	t_{CF}
Corre-lation Time	t_{inst}	ADC ADD ARPL BOUND BSWAP	*	*	*	*	*	*
		AND	*	*	*		*	*
		AAD AAM		*	*		*	
		AAA AAS				*		*
		BSF BSR			*			
		BT BTA BTC BTS						*

t_{inst} – instruction completed

t_{SF} – SF flag written	t_{OF} – OF flag written
t_{AF} – AF flag written	t_{ZF} – ZF flag written
t_{CF} – CF flag written	t_{PF} – PF flag written

Table 4-9 Arithmetic Flags/Instruction Coverage Model

In this model, the correlation time for all attributes is t_{inst}, when the corresponding instruction is completed. Five instructions — ADC, ADD, ARPL, BOUND and BSWAP — must be observed when each of the six arithmetic flags have the values zero and one after the instruction completes. The AND instruction interacts with all of the flags except AF. AAD and AAM are only to be recorded when the SF, ZF or PF flags are written. BSF and BSR are only recorded with ZF and BT, BTA, BTC and BTS are only recorded when CF is written.

Having examined the top-level design of several models, let's turn to the second step of coverage model design, detailed design.

4.4. Detailed Design

Detailed design concerns itself with mapping the coverage model design to the verification environment. In other words, how must the design be architected so that it may be implemented in ***e***?[9] In order to answer the question, three specific questions must be answered:

1. What must be sampled for the attribute values?
2. Where in the verification environment should we sample?
3. When should the data be sampled and correlated?

The answer to the first question maps verification environment fields and variables, or device signals and registers, to attributes. The answer to the second question determines where in the verification environment data sampling will be performed. The answer to the third question maps the sampling and correlation times to specific events in the environment. Each is addressed in

[9] The same detailed design process is required if the implementation language were Verilog, SystemVerilog, VHDL or SystemC.

the following sections.

4.4.1. What to Sample

The data to be captured for each of the attributes defined during top level design must be associated with a data source in the environment, either in the verification environment or in the device under verification. The verification interface of the coverage model determines, to a large extent, what data should be sampled. For example, the attributes of an input coverage model will be sampled from data injected into the device. The attributes of output coverage model will be sampled from data captured on device outputs. Some coverage models may require attributes that may only be captured from within the DUV or may compose attributes from all three sources. Let's address each of these sampling requirements in turn.

Input attributes should be sampled by a monitor that has access to the primary inputs to the device. The monitor should sample data from input signals at valid times specified by the device specification. The monitor should not retrieve its data from a stimulus generator because creating such a dependency would compromise the ability to reuse the monitor for recording coverage at a subsequent integration level. The following figure (4-7) depicts a monitor sampling data from a device's primary inputs.

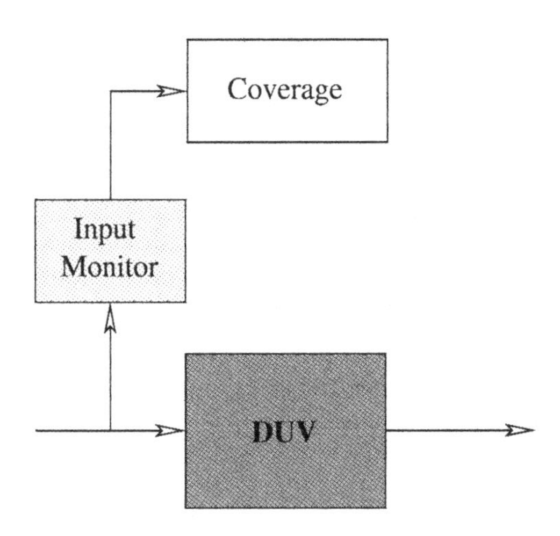

Figure 4-7 Sampling Input Attributes

The block labeled "Coverage" is a distinct aspect from the "Input Monitor" block because it has a particular responsibility whereas the monitor is more general-purpose. The coverage block is solely responsible for recording functional coverage data. The monitor block is a generic data capture facility, used by both coverage and checking aspects. It is implemented as either an extension to an ***e***RM agent[10] responsible for the device input interface or as a distinct ***e*** unit within the agent. In either case, there is a clean separation between its implementation and that of the monitor. The coverage aspect defines ***e*** coverage groups, their sampling events and any supporting data structures and procedural code.

An example of a sampled input attribute is the SMI# pin of an IA-32 processor. This asynchronous signal would be sampled when asserted. Since it is an active low signal (as indicated by the "#" signal name suffix), it would be captured when driven to the value zero.

Output attributes should also be sampled by a monitor but, of course, from the primary outputs of the device. The monitor should sample data at valid times specified by the device specification. As with device inputs, an ***e***RM agent should be written for each output interface, containing a monitor for passive signal sampling. The following figure (4-8) depicts a monitor sampling data from a device's primary outputs.

[10] An ***e***RM agent is a modular ***e*** program component responsible for interfacing to a device or one of its interfaces.

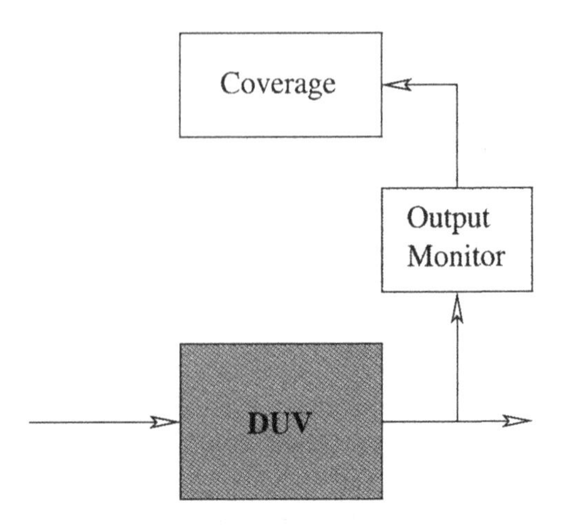

Figure 4-8 Sampling Output Attributes

Example of an output attribute to be sampled is the SMI acknowledge transaction. This transaction is the response of an IA-32 processor to an assertion of the SMI# pin. Since the transaction does not map directly to a device output, the monitor must recognize the transaction protocol on the bus and when the transaction completes, record it for the coverage model.

Device internal attributes also have an associated monitor, responsible for capturing internal signal and register values. Unlike the input and output interfaces of the device, internal signals are not as well specified (if they are specified at all). Sampling of such internal signals must be coordinated with the design team in order to minimize volatility.

Because internal signals may be quite volatile and are frequently unspecified, the use of internal attributes for coverage model design should be minimized. They impose an additional maintenance burden on the verification team, once implemented, because the monitor and coverage must continually track design changes. One strategy I have seen successfully employed to address this is defining a set of fixed signals and registers for verification use. The verification team identifies internal signals and registers required for coverage measurement and the design team agrees to minimize their changes. This "verification interface" is treated the same as an external device interface because it is specified and (nearly) frozen.

The following figure (4-9) depicts a monitor sampling data from the DUV's internal signals and registers.

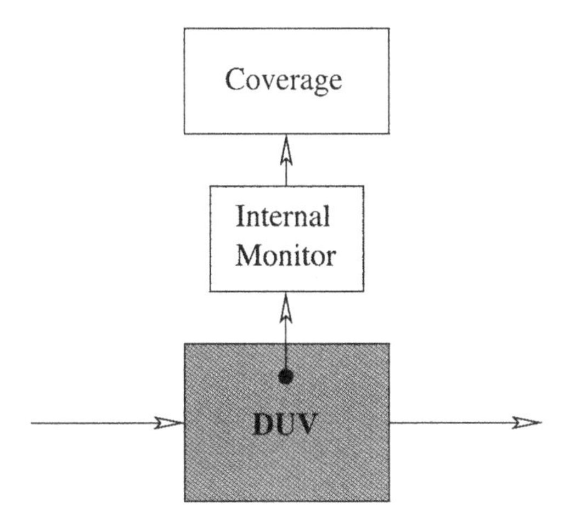

Figure 4-9 Sampling Internal Attributes

The second question to be answered during detailed design of a coverage model is *where in the verification environment should we sample?* This is addressed in the next section.

4.4.2. Where to Sample

A dynamic verification environment is composed of three aspects: coverage measurement, stimulus generation and response checking. In an ***e*** implementation, these aspects are distributed among three ***e*** language aspects of the environment.[11] Data sampling for coverage measurement is part of the coverage measurement aspect of the environment.

The coverage measurement aspect will be implemented in a set of files that extend existing objects of the verification environment. In an ***e***RM-compliant environment, the agents responsible for monitoring input, output and internal interfaces are extended. The coverage groups may either be defined in a subtype of the monitor in an agent or in their own unit.

[11] ***e*** is an aspect-oriented programming language. This means that a capability of the environment, whose implementation requires contributions from one or more unrelated objects, may be implemented in a single file which extends each of those objects. See http://aosd.net/ for more information on AOP (March 2004).

If a coverage group is defined in a monitor subtype, the subtype should be "has_coverage," per *e*RM guidelines. For example:[12]

```
extend has_coverage ap_input_agent_u {
  cover inst_completed using text =
    "Arithmetic Flags and Instructions" is ...
}
```

If the coverage group is defined in its own unit, the unit is declared in the agent. For example:

```
extend ap_input_agent_u {
  coverage : ap_input_cov_u is instance
};

unit ap_input_cov_u {
  cover inst_completed using text =
    "Arithmetic Flags and Instructions" is ...
}
```

Typically, if we are measuring output coverage, a monitor on an output interface of the device will capture data used by a data checker — such as a scoreboard — as well as by a coverage model. If attributes are required from device inputs, they should be captured from a monitor on the device inputs. It may be tempting to place input data capture in the generation aspect of the environment, but what happens when this aspect is reconfigured or removed in the verification environment of a subsequent integration of the device? Input coverage could no longer be measured on that interface.

The third question to be answered during detailed design of a coverage model is *when should the data be sampled and correlated?*

4.4.3. When to Sample and Correlate Attributes

During top-level design, we selected a sampling time for each attribute and a correlation time for each set of related attributes in the coverage model table. In this section, each sampling and correlation time is refined by defining it in terms of the verification environment.

[12] If the monitor is declared in the agent, the coverage group should be defined in the "has_coverage" subtype of the monitor unit.

The first coverage model investigated, "Abbreviated Virtual-8086 Mode Coverage Model," table 4-4, defines the sampling times "CR0 write" and "EFLAGS VM write" and the correlation time "CR0 or EFLAGS VM write." CR0 is written by the move-to-control-register instruction: MOV CR0, *r32*. We define an event, `move_to_CR0_e`, to be emitted by an internal monitor whenever a move-to-CR0 instruction completes.

EFLAGS.VM is written by either a task switch or return from a protected mode interrupt. Detecting either of these conditions from a black box perspective is rather complex so we again turn to an internal monitor. Another event, `eflags_vm_write_e`, is defined to be emitted by an internal monitor whenever the EFLAGS.VM bit is written.

The full correlation time for this model is defined by the "or" of these two events: `@move_to_CR0_e or @eflags_vm_write_e`.[13] The event `CR0_or_VM_write_e` is defined using this temporal expression:

```
event CR0_or_VM_write_e is
   @move_to_CR0_e or @eflags_vm_write_e
```

The sampling and correlation times for each of the earlier coverage models are refined in the same fashion. An observation interface is selected. A temporal expression is written to define the sampling or correlation time. Event operands are defined, as needed, to be emitted at constituent moments of the temporal expression.

Now, let's turn to the third and final step of coverage model development, implementation.

4.5. Model Implementation

If you've done a good job of designing a coverage model — top-level and detailed design — implementing the model should be less than 20% of the total effort. Why is that? Most of the difficult choices will have been made, those requiring analysis of the device specifications and ferreting out the precise intended relationships among device features. The remaining abstraction refinement of the model will be dictated by the structure of your verification environment and the semantics of the implementation language.

[13] "@*event*" is an ***e*** temporal expression that succeeds whenever *event* is emitted.

The verification environment is assumed to be structured into three primary aspects: coverage measurement, stimulus generation and response checking. The coverage measurement aspect will slice across a number of objects which implement the verification environment. The main objects to be extended will be monitors: input monitors, output monitors and internal device monitors. Let's look at the implementation of the GPR/arithmetic flags coverage model.

In the following implementation, the instruction set architecture (ISA) state of the CPU is maintained in the unit `ISA_state_u` and declared in agent `has_coverage ap_input_agent_u`.

```
extend has_coverage ap_input_agent_u {
  isa : ISA_state_u is instance
}
```

An ***e*** unit is special kind of struct that must be specified as either an instance ("`is instance`") or a reference. An ***e*** struct is like a C++ or Java class in that it includes data and procedural members, as well as temporal members. The data members are referred to as fields, two of which we are concerned with: `dest_gpr` and `eflags`. `dest_gpr` is the destination register of the last executed instruction. `eflags` is the current value of the EFLAGS register.

```
unit ISA_state_u {
  dest_gpr : gpr_t;
  eflags   : eflags_s
}
```

`dest_gpr` is of type `gpr_t`, a user-defined enumerated type:

```
type gpr_t : [
  EAX, EBX, ECX, EDX, ESI, EDI, EBP, ESP,
   AX,  BX,  CX,  DX,  SI,  DI,  BP,  SP,
   AH,  BH,  CH,  DH,
   AL,  BL,  CL,  DL
]
```

and `eflags` is an instance of struct `eflags_s`:[14]

[14] Since the word "of" is a reserved word in ***e***, I use "vf" instead for all references to the overflow flag in this code example.

```
struct eflags_s {
  cf : bit;
  pf : bit;
  af : bit;
  zf : bit;
  sf : bit;
  vf : bit
}
```

The coverage group is defined in the `has_coverage` subtype of the agent `ap_input_agent_u`:

```
extend has_coverage ap_input_agent_u {
  cover eflags_written_e is {
    item gpr : gpr_t = isa.dest_gpr;
    item cf  : bit   = isa.eflags.cf;
    item pf  : bit   = isa.eflags.pf;
    item af  : bit   = isa.eflags.af;
    item zf  : bit   = isa.eflags.zf;
    item sf  : bit   = isa.eflags.sf;
    item vf  : bit   = isa.eflags.vf;
    cross gpr, cf, pf, af, zf, sf, vf
  }
} // extend ap_input_agent_u //
```

The coverage group sampling event is emitted on the rising edge of the EFLAGS write signal. An ***e*** coverage group sampling event defines the correlation time of the attributes defined by its items. The sampling event must be defined in the base type for reference by a subtype.

```
extend ap_input_agent_u {
  event eflags_written_e
    is rise(sig_eflags_write$)
}
```

`sig_eflags_write` is the name of a single bit input port bound to the EFLAGS write signal.[15] `sig_eflags_write$` is a reference to its value.

[15] ***e*** ports are preferred over computed names (ex. '(sig_eflags_write)') because they support modularity and improve performance.

With the ISA state declared earlier, we now write the code to manage its values. First, there is the destination general purpose register code:

```
extend has_coverage ap_input_agent_u {
  event gpr_write_e is rise(sig_gpr_w$);

  on gpr_write_e {
    isa.dest_gpr = sig_dest_gpr$
  };
```

When event gpr_write_e is emitted, sig_dest_gpr$ is assigned to isa.dest_gpr. Next, we have the arithmetic flags management code. Each event defining an attribute sampling time is emitted when a flag is written.

```
  event cf_write_e is rise(sig_cf_w$);
  event pf_write_e is rise(sig_pf_w$);
  event af_write_e is rise(sig_af_w$);
  event zf_write_e is rise(sig_zf_w$);
  event sf_write_e is rise(sig_sf_w$);
  event vf_write_e is rise(sig_vf_w$);
```

Finally, the on-blocks capture each flag value when its corresponding event is emitted.

```
  on cf_write_e { isa.eflags.cf = sig_cf$ };
  on pf_write_e { isa.eflags.pf = sig_pf$ };
  on af_write_e { isa.eflags.af = sig_af$ };
  on zf_write_e { isa.eflags.zf = sig_zf$ };
  on sf_write_e { isa.eflags.sf = sig_sf$ };
  on vf_write_e { isa.eflags.vf = sig_vf$ }
} // extend ap_input_agent_u //
```

The GPR/arithmetic flags models has a matrix structure. How would we implement a hierarchical coverage model or hybrid model? A hierarchical coverage model may be implemented in two different ways: using subtyped coverage or per-instance coverage.

Subtyped coverage makes use of extending a coverage group under when subtypes of a base struct. Let's use the register pairs coverage model, described earlier, as an example. Its structure is reproduced below for reference (figure 4-10).

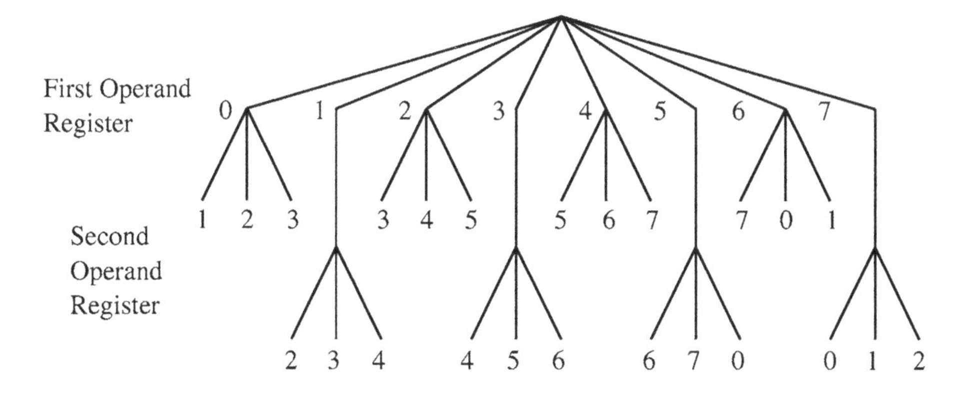

Figure 4-10 Register Operand Pairs Structure

Using subtyped coverage, this model would be implemented like this. First, the hosting struct `reg_pair` is declared in the agent:

```
extend has_coverage ap_reg_agent_u {
  reg_pair : reg_pair_s
};
```

The hosting struct is defined with a field for each register operand. The coverage group `register_pairs` is defined in the base struct with the first register operand `reg1`. The coverage group correlation event `register_pairs` is also declared.[16]

```
struct reg_pair_s {
  reg1 : reg_t;
  reg2 : reg_t;
  cover register_pairs is {
    item reg1
  };
  event register_pairs
};
```

The register type is an enumerated type.

```
type reg_t : [R0, R1, R2, R3, R4, R5, R6, R7];
```

The remaining code populates the coverage group with a `reg2` item for each value of `reg1`. For example, when `reg1` is `R0`, `reg2` is only recorded

[16] The code that emits the event is not shown.

when it has the value R1, R2 or R3. Unique names must be given to each item because the items share a common coverage group name space.

```
extend R0'reg1 reg_pair_s {
  cover register_pairs is also {
    item R0reg2 : reg_t = reg2
      using ignore = (reg2 not in [R1..R3])
  }
};

extend R1'reg1 reg_pair_s {
  cover register_pairs is also {
    item R1reg2 : reg_t = reg2
      using ignore = (reg2 not in [R2..R4])
  }
};

extend R2'reg1 reg_pair_s {
  cover register_pairs is also {
    item R2reg2 : reg_t = reg2
      using ignore = (reg2 not in [R3..R5])
  }
};

extend R3'reg1 reg_pair_s {
  cover register_pairs is also {
    item R3reg2 : reg_t = reg2
      using ignore = (reg2 not in [R4..R6])
  }
};

extend R4'reg1 reg_pair_s {
  cover register_pairs is also {
    item R4reg2 : reg_t = reg2
      using ignore = (reg2 not in [R5..R7])
  }
};
```

```
extend R5'reg1 reg_pair_s {
  cover register_pairs is also {
    item R5reg2 : reg_t = reg2
      using ignore = (reg2 not in [R6,R7,R0])
  }
};

extend R6'reg1 reg_pair_s {
  cover register_pairs is also {
    item R6reg2 : reg_t = reg2
      using ignore = (reg2 not in [R7,R0,R1])
  }
};

extend R7'reg1 reg_pair_s {
  cover register_pairs is also {
    item R7reg2 : reg_t = reg2
      using ignore = (reg2 not in [R0..R2])
  }
}
```

The same model may be implemented using per-instance coverage as follows. The register type is defined, as well as the coverage group correlation event, `register_pairs`.

```
type reg_t : [R0, R1, R2, R3, R4, R5, R6, R7];

extend ap_reg_agent_u {
  event register_pairs
};
```

The register fields for operands one and two are declared in the `has_coverage` subtype of the agent.

```
extend has_coverage ap_reg_agent_u {
  reg1 : reg_t;
  reg2 : reg_t;
```

The coverage group is defined in the agent with the first item, `reg1`, specified as `per_instance`. `per_instance` allows us to define different `reg2` restrictions for each value of `reg1`.

```
cover register_pairs is {
  item reg1 using per_instance
};
```

For example, the first instance, `(reg1 == R0)`, specifies that `reg2` should only be recorded when it has the value `R1`, `R2` or `R3`.

```
cover register_pairs(reg1 == R0) is also {
  item reg2 using ignore = (reg2 not in [R1..R3])
};

cover register_pairs(reg1 == R1) is also {
  item reg2 using ignore = (reg2 not in [R2..R4])
};

cover register_pairs(reg1 == R2) is also {
  item reg2 using ignore = (reg2 not in [R3..R5])
};

cover register_pairs(reg1 == R3) is also {
  item reg2 using ignore = (reg2 not in [R4..R6])
};

cover register_pairs(reg1 == R4) is also {
  item reg2 using ignore = (reg2 not in [R5..R7])
};

cover register_pairs(reg1 == R5) is also {
  item reg2 using ignore = (reg2 not in [R6,R7,R0])
};

cover register_pairs(reg1 == R6) is also {
  item reg2 using ignore = (reg2 not in [R7,R0,R1])
};

cover register_pairs(reg1 == R7) is also {
  item reg2 using ignore = (reg2 not in [R0..R2])
}
} //extend has_coverage ap_reg_agent_u//
```

One advantage of implementing a hierarchical model using per-instance coverage is immediately apparent. It is usually much more compact than subtyped coverage. The reason is that the necessary attribute hierarchy may be constructed of coverage subgroups ("`cover register_pairs (reg1 == R6) is` ..."). The subgroup mimics the inheritance hierarchy of *e* structs and units. For the same reason, this application of per-instance also has a drawback. Although the struct/unit hierarchy reflects the object-oriented design of a verification environment, coverage subgroups are completely orthogonal to it.

A hybrid model would be implemented using a combination of the matrix and hierarchical techniques shown above.

Summarizing the model implementation process, the "what," "where" and "when" questions addressed by detailed design are realized. Verification environment fields and device signals are written to capture attributes. *e*RM agents are identified to host the coverage groups, either in `has_coverage` subtypes or in dedicated units. Sampling and correlation times are implemented as temporal expressions bound to events.

4.6. Related Functional Coverage

In this section I briefly discuss three kinds of functional coverage deserving further investigation and application. They are finite state machine (FSM) coverage, temporal coverage and static verification coverage.

4.6.1. Finite State Machine Coverage

Automatic FSM extraction by code coverage programs is a relatively new capability.[17] Early code coverage tools defined pragmas for the designer to indicate to the code coverage tool an FSM and the structure used to implement it. Current tools recognize most FSM implementations and record state and arc coverage.[18] However, there are times when you may need or want to record FSM coverage in a functional coverage model. One situation is where your code coverage program does not recognize or extract a particular FSM. Another is when you need to interrelate the behavior of two state machines. Yet another is the need to record FSM behavior along with other functional coverage information.

[17] Circa 1999.

[18] See chapter 5, "Code Coverage."

In each case, a functional coverage model may be designed whose attributes include the FSM state variable and controlling next-state equation terms. The attribute values of the state variable are its defined states. Those of the next-state equation terms are their specified values. The behavior of the FSM may reflect any coverage model structure but hierarchical is the most common. The detailed design and implementation of this model follow the procedure outlined earlier in the chapter.

4.6.2. Temporal Coverage

The behavior of a DUV, as observed from any verification interface, usually has both data and temporal components. In other words, the boundary conditions exist in both the data domain and the time domain. The data domain is most familiar to the verification engineer. It comprises the world of values observed on buses and latched in registers. The time domain, on the other hand, is most often relegated to second class status. Moments define when data should be captured and intervals define how long data is valid. However, the sequential behavior of the device is given less attention than it deserves.

Quite often, devices are optimized for frequently repeated operations. They take advantage of data and time locality to deliver higher performance. Caches and translation look-aside buffers are good examples. Experience indicates this optimization logic is often rife with bugs. More so, half of these kinds of bugs result in no functional misbehavior. Instead, they lead to performance degradation or excessive power consumption. All this encourages us to verify temporal behavior and that includes recording temporal coverage.

Sequential verification is an area of active research and some internal tools[19] have been developed and employed. However, I am aware of no commercial tools that directly measure temporal coverage, nor offer implementation facilities for directly specifying temporal coverage models. Therefore, temporal behavior must be mapped into the data domain. Suppose the sequential relationship "Z occurs within 5 to 10 cycles after either X or Y occurs," written in the ***e*** temporal language as:

[19] "Piparazzi: A Test Program Generator for Micro-architecture Flow Verification" by Allon Adir, Eyal Bin, Ofer Peled and Avi Ziv, IBM Research Lab in Haifa, Israel, November 2003, Eighth Annual IEEE International Workshop on High Level Design Validation and Test.

```
{(@X or @Y); [4..9]; @Z}
```

is an important condition to observe as a coverage measurement. If the temporal expression (TE) is associated with an event:

```
event ev1 is {(@X or @Y); [4..9]; @Z} @clk
```

commercial tools do report the number of times the event was emitted. However, the elastic time interval `[4..9]` is not recorded. The TE may be instrumented to do so:

```
t        : time;
interval : time;

event ev1 is {
  (@X or @Y) exec {t = time};
  [4..9]     exec {interval = time - t};
  @Z
} @clk
```

and coverage of the field `interval` recorded. This is one way to map values from the time domain to the data domain for coverage measurement using existing commercial tools.

4.6.3. Static Verification Coverage

More often than not, verification teams are employing both static and dynamic methods in their quest to expose bugs. Coverage-driven verification requires engineers to define coverage goals and implement a verification environment to reach them. When the environment includes a formal tool, such as a model checker, what coverage is provided by that flow? In particular, what functional coverage is delivered by proven properties?

Each property has an associated semantic description, similar to the semantic description that leads the functional coverage model design process. The property descriptions may be compared with those of the developed coverage models. A property whose description is a superset of the description of a particular coverage model may be determined to have fully traversed the model, once the property is proven. If the property description is a subset of a coverage model, analysis is required to quantify what subset of the model may be considered "observed" when the property is proven. In either case, the rigorous nature of a proof and the quantitative nature of functional coverage are compromised in order to associate one with the other.

Another approach to quantify verification coverage delivered by a property checker has been used by Averant.[20] The properties proven by their static verification tool, Solidify, may be recorded as coverage points and reported by Specman, Verisity's verification automation product. However, it appears as though each property is represented by a single coverage point, offering little visibility into the nuances of the property terms. This would be equivalent to a low fidelity coverage model, unable to distinguish one variant of a behavior from another in recorded results.

There are many avenues to explore to bridge static verification methods with coverage measurement techniques. These will lead to the ability to quantify the functional, code and assertion coverage delivered by each proven property. Since the purpose of measuring and analyzing coverage is to determine verification progress and formal methods are a valuable tool for verification, we must discover a practical means for assessing the contribution of proofs on the coverage scales.

4.7. Summary

In this chapter, you learned how to use functional coverage to model device behavior at various verification interfaces. After an overview of the model design process, using a simple device as an example, the process of top-level design, detailed design and implementation of functional coverage models was described. As part of top-level design, selecting a coverage model structure that reflects the attribute relationships of the device is important choice. The three model structures — matrix, hierarchical and hybrid — were illustrated. Several coverage model implementations were examined in order to understand the implication of model structure on implementation. Lastly, I briefly reviewed three kinds of related functional coverage: FSM coverage, temporal coverage and static verification coverage.

[20] "Averant Announces Significant New Improvements in Solidify 2.6," June 10, 2002, DAC 2002 press release, http://www.averant.com/-release26.htm.

5. Code Coverage

In this chapter I explain the purpose of code coverage — sometimes referred to as structural coverage — its various metrics and a suggested use model.

Code coverage, as you'll recall from chapter 3, "Measuring Verification Coverage," defines an implicit implementation coverage space. An implicit implementation coverage space is one in which the coverage metrics are defined by the source we are observing and extracted from the device implementation. Unlike functional coverage, for which the attributes to be measured must be defined and their organization designed, code coverage attributes are predefined by the RTL and, hence, by the code coverage program.

However, an interesting trend in the capabilities of code coverage tools is apparent. Over time, metrics which had to be manually measured using functional coverage techniques have been integrated into code coverage tools. For example, only recently has finite state machine (FSM) extraction become widely available. Before then, the engineer interested in measuring FSM coverage had to implement the code by hand by either instrumenting the RTL or building an external coverage monitor.

Before diving into the various code coverage metrics, let's examine instance and module coverage.

5.1. Instance and Module Coverage

The implementation of a device in a given HDL includes modules composed of logic, registers, wires and events. These modules are instantiated one or more times, depending upon the amount of replicated logic. Code coverage may be measured and reported for the module definitions and for each instance of module components.

Module level coverage is appropriate for blocks that are replicated in a symmetric fashion. For example, if an operation may be parallelized through decomposition into a number of data paths, generally independent of one another, there is no reason to record coverage on a per-instance basis. When the metrics discussed below are recorded at the module level, the simulation performance degradation is reduced and disk and memory storage conserved.

Instance level coverage is required when a logic block is replicated, but distinct modes of operation are enabled for each instance. For example, a data encoder may be able to encode using two variants of the same encoding algorithm. The encoder itself may be duplicated — hard-wired to a configuration — in order to overlap encoding two data blocks, each using an alternate algorithm. In this case, distinct lines, statements, expressions and control paths will be exercised in each configuration. Instance level coverage must be used to gain the necessary visibility into the module.

5.2. Code Coverage Metrics

Code coverage metrics are implicit coverage attributes of the hardware description language, which are measured in the RTL device implementation. In the following sections, I review the most common metrics: line, statement, branch, condition, event, toggle and FSM coverage.

All code coverage programs provide a means for specifying a *hit threshold*. The hit threshold is the minimum number of times a metric must be observed in order to be counted as covered. This is discussed further in the use model section 5.3.2, "Record Metrics."

5.2.1. Line Coverage

The line coverage metric reports which lines of the RTL have and have not been executed. Full line coverage is defined as all non-comment lines executed a threshold number of times. This threshold is often user-specified. This is an example of VHDL line coverage:

Count	Source Line
1	`architecture arch of pipeline is begin`
1	`  request_bus_p: process begin`
57	`    prequest_n <= '1';`
57	`    penable_n  <= '1';`
57	`    wait until trst_n'event and trst_n = '0';`
57	`    wait until trst_n'event and trst_n = '1';`
57	`    prequest_n <= '0';`
57	`    wait until clk32'event and clk32 = '0'`
57	`      and pgrantin_n = '0';`
57	`    wait until clk32'event and clk32 = '0';`
57	`    penable_n <= '0';`
57	`    wait;`
1	`  end process;`
1	`end arch;`

A related coverage metric reported by some tools, block coverage, defines a sequence of statements with no branches to be a block. (A branch is introduced by any control flow statement such as an "if" or "while" statement.) If the first statement of the block is executed, all of the statements in the block are executed. Although block coverage offers the same visibility into RTL execution as line coverage, less overhead is introduced during simulation because only one counter is inserted per block.[1]

5.2.2. Statement Coverage

Statement coverage reports which RTL statements have and have not been executed. This metric is more precise than line coverage because statements may span multiple lines and more than one statement may occupy a single line. This is an example of Verilog statement coverage:

[1] The process of using code coverage is discussed in section 5.3, "Use Model."

Count	Source Line
1	`always @(ALUSelA or ReadData1 or PC) begin`
172	`  if (ALUSelA) begin`
36	`    MuxA = ReadData1;`
	`  end else begin`
136	`    MuxA = PC;`
	`  end`
	`end`

5.2.3. Branch Coverage

The branch coverage metric reports the count of control flow transfers for *if*, *case*, *while*, *repeat*, *forever*, *for* and *loop* statements. A control flow transfer interrupts the normal sequential execution of statements. Some code coverage tools refer to branch coverage as *arc coverage*, although this term is more often applied to FSM coverage, described later in this chapter. This is a typical example of reported branch coverage in VHDL:

```
Line  Source
 17   case int_ack_src_acc_reg is
        ...
 55     when "00" =>
 56       for i in 0 to fifo_size - 1 loop
 57         if i = c then
 58           update(i) <= '1';
 59           load(i)   <= '1';
 60         else
 61           update(i) <= '0';
 62           load(i)   <= '0';
 63         end if;
 64       end loop;
 65       c_inc <= 1;

 67     when "01" =>
 68       for i in 0 to fifo_size - 1 loop
 69         update(i) <= '1';
 70         if i = c_minus_2 then
```

Line	Source
71	`load(i) <= '1';`
72	`else`
73	`load(i) <= '0';`
74	`end if;`
75	`end loop;`
76	`c_inc <= -1;`
	...
99	`end case;`

From Statement	To Statement	Branch Count
17	55	6
17	67	15
17	99	1
57	58	4
57	60	2
70	71	7
70	72	8

Table 5-1 Branch Coverage

A related, yet more complete control flow metric is *path coverage.* Path coverage reports the number of times every possible path through the code was executed. For example, an if/else statement followed by a second if/else statement has four paths:

1. if 1 true, if 2 true
2. if 1 true, if 2 false
3. if 1 false, if 2 true
4. if 1 false, if 2 false

Specific to loop constructs, *loop coverage* reports a count of how often each loop in the code is executed a specific number of times, usually zero, one or two times. This is useful for determining how well loops were exercised for their iteration boundary conditions.

5.2.4. Condition Coverage

Condition coverage records the number of times each permutation of the terms of a Boolean expression cause the complete expression to evaluate to true or false.[2] Consider the expression (A && B) || C || D. This expression evaluates to true under three conditions:

1. A && B
2. C
3. D

and to false under two conditions:

1. !A && !C && !D
2. !B && !C && !D

A stricter form of condition coverage — exclusive condition coverage — is sometimes available that requires each term to be the controlling term (i.e. sole reason) for an expression to evaluate true or false. For the example above, the expression evaluates to true for these five exclusive conditions. The controlling term is in boldface below:

1. **A && B** && !C && !D
2. !**A** && B && **C** && !D
3. A && !**B** && **C** && !D
4. !**A** && B && !C && **D**
5. A && !**B** && !C && **D**

5.2.5. Event Coverage

The event coverage metric records the number of times an event is triggered, occurs or is emitted.[3] In Verilog and SystemVerilog, an unnamed event is triggered when either the value of a wire or register is changed. A named event may be defined and the event explicitly triggered using the -> operator.

[2] Software developers refer to this as *modified condition decision coverage* (MCDC).

[3] Events are referred to as *triggered* in Verilog and SystemVerilog; *occurred* on a signal in VHDL and *emitted* in ***e***.

In VHDL, an event occurs on a signal if the value of that signal changes.

5.2.6. Toggle Coverage

Toggle coverage reports the number of times each bit of a register or wire has toggled its value. For example, the toggle coverage reported for the following register:

```
reg [10:0] state;
```

may be reported as:

Bit	0 → 1	1 → 0
reg[10]	12	11
reg [9]	5	5
reg [8]	9	10
reg [7]	31	30
reg [6]	31	30
reg [5]	31	30
reg [4]	31	30
reg [3]	31	30
reg [2]	31	30
reg [1]	31	30
reg [0]	31	30

Table 5-2 Toggle Coverage

Some code coverage tools report not only zero-to-one and one-to-zero transitions, but also transitions to and from undefined ("X") and tristate ("Z").

5.2.7. Finite State Machine Coverage

Modern code coverage tools identify and extract finite state machines from RTL. There are several FSM metrics of interest to the verification engineer. The most basic is state visitation: how many times was each state of a state machine entered? Another is arc coverage: how many times did the FSM transition from one state to each of its neighboring states? Arc coverage should be reported for the subexpressions of each next state equation as we saw for condition coverage. A third FSM metric is sequential arc coverage, often called transition coverage. Sequential arc coverage identifies state visitation sequences of various lengths and records the number of times each sequence was traversed.

Consider the following FSM composed of five states and nine arcs. Next to each arc is its next-state equation.

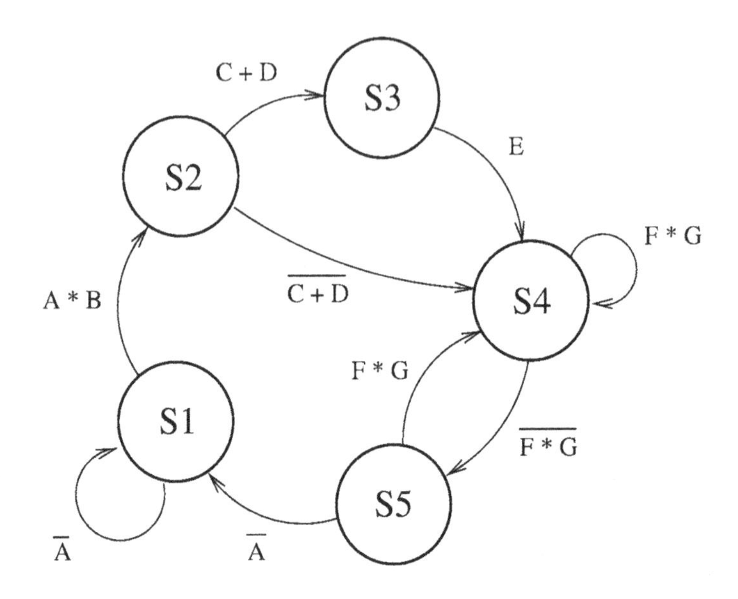

Figure 5-1 Finite State Machine

State coverage for this FSM might be reported as follows:

State	Visits
S1	315
S2	45
S3	32
S4	211
S5	45

Table 5-3 FSM State Coverage

while arc coverage might be reported as:

Arc	Next-State Equation	Next-State Equation Terms	Traversals
S1→S1	$\bar{A}$	A	239
S1→S2	A * B	A, B	45
S2→S3	C + D	C, D	32
		C	20
		D	12
S2→S4	$\overline{C+D}$	C, D	117
		C	53
		D	64
S3→S4	E	E	32
S4→S4	F * G	F, G	45
S4→S5	$\overline{F*G}$	F, G	45
		F	3
		G	42
S5→S4	F * G	F, G	7
S5→S1	$\bar{A}$	A	76

Table 5-4 FSM Arc Coverage

Lastly, sequential arc coverage — for two-arc transitions beginning at state S1 — might be reported as:

Arc Sequence	Traversals
S1→S1→S1	239
S1→S1→S2	15
S1→S2→S3	32
S1→S2→S4	117

Table 5-5 FSM Sequential Arc Coverage

Before moving to the use model for code coverage, I address a way to improve code coverage fidelity by recognizing that internally observed metrics are more valuable if propagated to a checking interface. This is controlled *and* observed coverage.

5.2.8. Controlled and Observed Coverage

In the previous sections I discussed the coverage metrics commonly available from code coverage programs. Each metric was considered within the context of its associated RTL, with no consideration for the effect a recorded metric had on the running simulation. We may observe the execution of a statement, perhaps an assignment statement. What if the assigned value never propagates to an observation interface for checking? What if that statement is only executed one time in all of the regression runs and the value is never propagated? A potentially erroneous assignment is never checked, but recorded as observed. This situation bit me one time during the verification of a complex design.

I was responsible for verifying a superscalar, 3-issue IA-32 microprocessor. Since I was unable to convince management to invest in the development or purchase of an automated generation environment, my verification team had manually written hundreds of assembly language tests to verify the instruction set architecture (ISA). Nevertheless, we employed a coverage-driven verification methodology[4] to achieve 100% coverage on our ISA functional coverage models. Despite employing a rigorous CDV methodology, in first silicon we discovered a bug in one of the branch instructions. I knew we had measured 100% branch coverage and the self-checking branch tests had been thoroughly reviewed. How did this bug escape?

Two groups of tests we had developed were the floating point tests and the branch instruction tests. Unbeknown to us, the single occurrence of a particular branch instruction scenario was coded in one of the floating point tests. The floating point test, not responsible for branch instruction verification, was insensitive to the misbehavior of the branch instruction. The branch instruction test suite mistakenly excluded this particular branch scenario. When we functionally graded[5] the full test suite, there were no reported branch coverage holes. At the same time, all tests — including the branch tests — passed without error.

The problem was now apparent: we had not restricted the measurement of branch instruction functional coverage to the branch tests. More generally,

[4] This was 1993.

[5] The term "functionally grade" means to measure the functional coverage of a set of simulations or tests.

coverage was measured in one set of stimuli but not correlated to response checking in the same stimuli. Branch coverage *controlled* test execution but it was not *observed* by branch behavior checking code. The lesson for functional coverage is to make sure each functional coverage model has corresponding — and concurrently active — data and temporal checking code. How do we address the problem when using code coverage?

The code coverage program must provide a means of sensitizing the recording of code coverage metrics to a defined observation interface, generally a monitor of the verification environment checking aspect. For each metric — line, statement, branch, condition, event, toggle and FSM — the program must be able to trace the effect of the observed metric to a selected observation point through the RTL hierarchy. As of the publication date of this book, I am only aware of one commercial program that supports this feature.[6] Hopefully, in the near future other vendors will incorporate this capability into their offerings.

Having explored the benefit of correlating control and observation, how do we use code coverage? In other words, what is the recommended use model?

5.3. Use Model

Code coverage reports how well the RTL implementation of the device has been exercised from the perspective of each of the metrics discussed earlier. Since the RTL is quite volatile during the early stages of design, our interest in how well it has been exercised peaks later in the design cycle.[7] Let's walk through each step of the code coverage process: instrumentation, metric recording and analysis.

5.3.1. Instrument Code

The first step for using code coverage is to instrument the RTL. Select the code modules, hierarchies or instances you want to observe. Next, select the metrics you want to record. The amount of code instrumented and number of metrics measured will determine how much your simulation rate is

[6] "Synopsys VCS' Observed Coverage Technology," 2002, http://www.synopsys.com/products/simulation/obc_wp.pdf (March 2004).

[7] See figure 7-5, "Code Coverage Use," in chapter 7, "Coverage-Driven Verification."

degraded. Restrict these to what is required to meet your coverage goals.

The last instrumentation step depends upon the particular tool you are using. Some tools require no further action before simulation to begin recording metrics. Others require an instrumentation/compilation step wherein they insert code into the RTL which defines counters and increments them. Consult your code coverage tool documentation for the details on creating an instrumented, ready-to-simulate design database.

5.3.2. Record Metrics

The second step in the application of code coverage is recording metrics. The actual recording of metrics is performed by the simulator during simulation. However, the recorded data needs to be organized for subsequent analysis.

Each of the recorded metrics has an associated threshold or hit count. The default value is usually one. In high-risk areas of the RTL, with perhaps an unusual amount of complexity, you should consider increasing the threshold for the metrics recorded in these areas. This "over-sampling" will mitigate the risk of observing a statement, subexpression or FSM arc at a time the device is relatively quiescent.

Each simulation is usually identified by one or more unique identifiers. If an autonomous verification environment is used, the only attribute distinguishing one simulation from the next is the random seed. If a test-driven environment is used, a test number or id, in addition to a possible random seed, identifies a simulation. For a given snapshot of the RTL, code coverage should be accumulated for all of its simulations. After code coverage has been measured for an RTL release, it must be analyzed.

5.3.3. Analyze Measurements

The third step in using code coverage is analyzing the measurements. We need to exclude from analysis irrelevant data and focus on the meaning of recorded metrics.

Before examining how to interpret recorded metrics, keep in mind that code coverage cannot reveal necessary, but missing, RTL. If RTL required to implement a device requirement has not been written, it can only be identified by functional or assertion coverage, not by code coverage.

Often, you may be overwhelmed with the amount of information presented by the coverage tool. Data filtering addresses this, the topic discussed in the next section.

5.3.3.1. Filtering Measurements

When code coverage is first enabled and metrics selected, the amount of reported information may be daunting. The reported results must be filtered to exclude known illegal conditions, known unused logic and low-priority code coverage holes.

Known illegal conditions are "else" clauses of "if" statements and unused case default statements for conditions that should not happen. The designer may have logic to handle the expected values zero, one and two from a 2-bit bus. If the bus returns the value three, a default case or "else" clause if often included to report an error through a displayed message. This logic is not expected to be executed, so it should be filtered. A good code coverage tool allows filtering at the level of most of the metrics: line, statement, FSM state, etc. Some provide pragmas that may be inserted in the RTL for the designer to tag logic as unused.

Another reason for filtering results is known unused logic. A module may be used twice in the design, but configured differently for each instance. The configuration often deactivates a subset of the module logic, preventing it from ever being used in a given instance. That logic should be tagged as unused, either in the RTL or using the user interface or configuration file(s) of the code coverage program.

Unless expected unused (or unexercised) logic is excluded from reported results, the code coverage measurements will be misleadingly low. Take the time to construct filter files and instrument the RTL with pragmas before wading through these coverage results.

Finally, there may be coverage holes you choose not to fill. These coverage holes correspond to RTL that is rarely exercised or low-risk or both. If a particular set of metrics are only observed under rare operational conditions and these conditions are also difficult to reproduce during verification, you may choose to filter them. Also, if a coverage hole corresponds to RTL with little, if any, complexity, you may also choose to spend scarce resources filling higher risk holes.

Now, let's turn to analyzing each of the code coverage metrics.

5.3.3.2. Line Coverage Analysis

The first metric to be analyzed, line coverage, indicates which RTL lines have been executed. A line may contain a partial statement, a full statement or multiple statements. If a line contains a partial statement, it must be examined along with the other lines which define it. If a line contains a complete statement but the line is reported as not executed, there are a couple of possible reasons.

One possibility is that the line cannot be executed because the data and control flow of the code prevent it. Another is that the condition required to execute the line is rare but it has not yet been created. If an autonomous verification environment is used, the probability of creating this condition may need to be increased. If a directed tests are driving the environment, they may not contain the necessary stimuli to create the condition.

Returning to the third line/statement relationship, multiple statements may be defined on a single line. If so and the line is reported as not executed, the same analysis used for a single statement on a line should be applied to each statement.

5.3.3.3. Statement Coverage Analysis

Statement coverage reports which RTL statements have and have not been executed. For each statement which has not been executed, the analysis described above for statements in lines should be applied.

5.3.3.4. Branch Coverage Analysis

Branch coverage reports control flow transfers between RTL statements. Let's consider each type of branch statement in turn.

An *if* statement has two possible branches: the *if* (true condition) branch and an optional *else* (false condition) branch. If either branch has not been executed, the *if* conditional expression should be examined to determine if the expression may, in fact, assume both true and false values.

The *case* statement is quite analogous to the *if* statement, except that it may have more than two branches. If one or more branches are reported as not executed, the case statement expression must be examined to make sure it may evaluate to each of the case label values.

The *while*, *repeat*, *forever*, *for* and *loop* statements are each characterized by a reverse control flow transfer, either conditional or unconditional. The *while* statement has three branches: one bypassing the body of the loop when the *while* condition is false, a second sequential transfer from the *while* test into the body and a third unconditional branch from the end of the body back to the conditional expression. If either of the first two branches are never taken, the *while* condition must be examined to make sure both the first and subsequent evaluations of the expression may evaluate to true and false. The third unconditional reverse branch must be observed if the sequential transfer into the loop body is recorded.

The Verilog/SystemVerilog *repeat* statement also has three branches so its statement coverage should analyzed in the same manner as the *while* statement.

The Verilog/SystemVerilog *forever* statement is not very interesting because, if the statement itself is encountered, both of its branches (sequential transfer into loop body and reverse transfer to top of body) will always be executed.

The Verilog/SystemVerilog *for* statement is essentially a repackaged *while* loop in which the initialization assignment and iteration variable increment are specified in the *for* expression itself. It should be analyzed just like the *while* statement.

The VHDL *loop* statement is used to construct an unconditional loop, *while* loop or *for* loop. These are each analyzed like the corresponding Verilog loop statements.

5.3.3.5. Condition Coverage Analysis

Condition coverage records how extensively the terms of expressions have been evaluated. Considering the example we used earlier, (A && B) || C || D, it has four Boolean terms: A, B, C and D. If the coverage program reports that any of these terms have not been observed evaluating to true or false, the constituent signal and register operands from which it is composed must be examined to determine why this is the case.

For example, a mutually exclusive relationship between A and B (A = !B) would prevent the term (A && B) from ever being true. Likewise, if C is defined to always be the inverted value of D (C = !D), the two false conditions will always be observed, independent of the values of A, B, C and D:

1. (!A && !C && !D) = (!A && !C && !(!C)) = (!A && !C && C) = (!A && 0) = 0
2. (!B && !C && !D) = (!B && !C && !(!C)) = (!B && !C && C) = (!B && 0) = 0

5.3.3.6. Event Coverage Analysis

Event coverage reports observed and unobserved moments in time. That is, if the conditions required to cause an event to be triggered never occur, the event will not be triggered. For each event which is not observed, the conditions to be met must be examined and compared against the simulations whose event coverage was measured.

Typically, an event is associated with a signal that has a cone of logic determining its value. In Verilog and SystemVerilog, if a signal has a static value throughout the simulation, no associated unnamed event will be triggered. (The same is true of a VHDL event on a signal.) The signal may be static because it was mistakenly OR-ed with an intended constant signal, such as a configuration pin. A named event may never be recorded because the -> operator in a statement was never executed. Examining the corresponding line or statement coverage should shed more light on the cause.

5.3.3.7. Toggle Coverage Analysis

Toggle coverage records transitions between values of bits in registers and wires. While the toggle coverage of a data register may yield little value, toggle coverage of a one-hot mux select bus will tell us whether or not all of the mux paths have been exercised. Toggle coverage of an address bus or other, control-oriented bus, will indicate whether or not basic activity was observed.

In general, toggle coverage serves as a general "liveness" or activity indicator. It provides a very coarse view of signal activity but associates no semantic meaning to recorded results.

5.3.3.8. FSM Coverage Analysis

FSM coverage records visits to states and single and sequential arc traversals. Assuming an FSM has been optimized to reduce its state count to a minimum, each state must be visited at least once. Likewise, each arc must be traversed at least one time and, in addition, it must be traversed for each controlling term of its next-state equation. This means that full next-state condition coverage is necessary for full FSM coverage.

The matter of sequential arc coverage — two, three or more arc traversal permutations — is more complex. In order to understand the necessary sequential arc depth and which arc permutations must be observed, it is necessary to understand what specified functionality and implemented device logic is only exercised under particular sequential scenarios. This understanding is only acquired through detailed analysis of the functional and design specifications and the RTL.

If you were to choose to capture FSM behavior using a functional coverage model, you would have to invest in the same analysis to design the model. However, you would be relieved of the back-end analysis required of code coverage FSM extraction.

5.4. Summary

In this chapter I explained that code coverage defines an implicit implementation coverage space, one defined by the RTL language, whose metrics are extracted from the device implementation. The distinction between module and instance coverage, and when to measure each, was explained. Each of the code coverage metrics — line, statement, branch, condition, event, toggle and FSM coverage — were explained and examples presented to illustrate their use. I introduced the topic of controlled and observed coverage, correlating the decision to record a coverage metric with observing its effect at a checking interface. A use model for code coverage was explained, starting with filtering results. Lastly, I discussed how to analyze each of the coverage metrics.

6. Assertion Coverage

In this chapter I introduce assertions and assertion coverage. I discuss several ways of classifying assertions, the kinds of properties that may be specified and the various property specification languages. The different meanings of "assertion coverage" are distinguished from one another. Finally, you will learn how to measure assertion coverage and analyze the results.

6.1. What Are Assertions?

Of the tools available to the engineer for capturing design intent, one excels at capturing intent late in the design process, at one of the lowest design abstraction levels: the assertion. As described in chapter 2, "Functional Verification," throughout the design process higher abstraction design intent is preserved while more detailed, lower abstraction design intent is added. The more detailed design intent is first captured in the design specification[1] and later, if assertions are used, in the RTL itself. Once embedded in the RTL, the device behavior is compared to its intended behavior each time the device is simulated. The assertions may also be formally proved through static analysis, obviating the need to verify the associated logic through simulation. With this understanding of how assertions are used, what exactly is an assertion?

As defined in chapter 1, "The Language of Coverage," an assertion is an expression that states a safety or liveness property. Safety or liveness is the first of three ways I classify an assertion. A safety property is a statement that something should never happen, so called because it states that an *unexpected* event should not happen. One example of a safety property is "No

[1] The functional specification defines high-level, black-box device requirements while the design specification defines its microarchitecture.

more than one select line of a one-hot select should be asserted at a time." Another example is "No more than one bit should change on a gray-code-encoded control bus per cycle."

A liveness property is a statement that something should eventually happen. By "liveness," I mean that an *expected* event must occur — i.e. the device must eventually exhibit activity. An example of a liveness property is "Grant must be asserted one to four cycles following the assertion of request." Another is "After reset is deasserted, eventually an instruction fetch cycle must be initiated."

A second way to classify an assertion is by its purpose: checking or coverage. The purpose of a checking assertion is to capture a requirement from either the design specification or the designer at the RTL abstraction level. It is one means of implementing the checking aspect of a verification environment.[2] The checking aspect may also be implemented in an HLVL verification environment.

The purpose of a coverage assertion is to report the occurrence of an expected event. It is one of a number of means of implementing a functional coverage model at the RTL abstraction level. Another is using an HLVL to write a low-level coverage aspect.

A third way to classify assertions is whether an assertion is concurrent or procedural. A concurrent assertion, sometimes called a declarative assertion, is evaluated whenever an event occurs, such as at the rising edge of a clock. A procedural assertion is evaluated whenever it is executed by an RTL execution thread. We can visualize the three dimensions of assertion classification as in figure 6-1 below:

[2] See Figure 2-2, "Functional Verification Aspects."

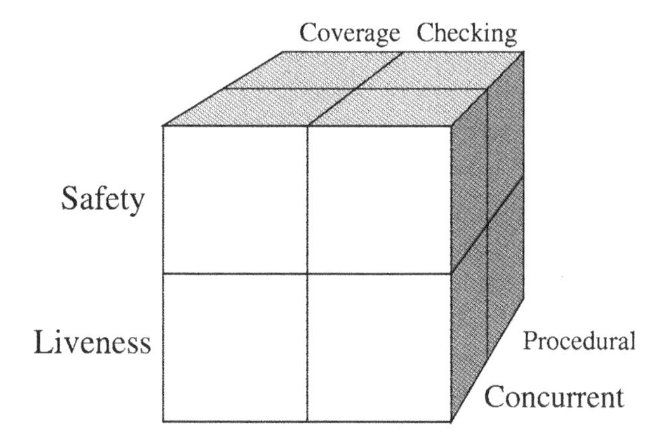

Figure 6-1 Assertion Classification

The most common assertion languages in use today are the Open Verification Library (OVL),[3] OpenVera Assertions (OVA),[4] Property Specification Language (PSL)[5] and SystemVerilog Assertions (SVA).[6] The Open Verification Library was invented by Harry Foster and Lionel Bening and later donated to the Accellera standards organization by Verplex Systems (subsequently acquired by Cadence Design Systems). OVL implements declarative assertions in Verilog as modules instantiated in the design RTL.

For example, the *assert_one_hot* assertion monitors its test expression at the positive edge of its triggering event or clock. If more than one bit has the value one when sampled, the assertion fails. The syntax of the assertion is:

```
assert_one_hot [ #( [severity_level]
  [, width, options] [, msg] ) ] <instance_name>
  (clk, reset_n, test_expr );
```

The bracket-delimited parameters are optional while the parenthesized argu-

[3] http://www.verificationlib.org/

[4] http://www.open-vera.com/

[5] http://www.accellera.org/

[6] "SystemVerilog 3.1: Accellera's Extensions to Verilog," http://www.accellera.org/ and http://www.eda.org/sv/.

ments are not. This is an example of its use in a Verilog module:

```
assert_one_hot #(1, 8) ovl23 (clk, reset_n, xsel);
```

When reset_n is deasserted, on each clk the 8-bit select line xsel is verified to have only one bit asserted.

OpenVera Assertions are derived from the Intel ForSpec temporal language. The OVA language is decomposed into five levels. The first is the bindings level, used to define the scope of the assertions. The second is the units level that defines the assertion name, port list and the sampling time. The third is the directives level, used to specify the properties to be monitored or checked. The fourth is the Boolean expressions level and the fifth is the event expressions level, used to define temporal sequences.

The following is a simple OVA assertion that detects an overflow of a counter implemented in Verilog module counter_13bit.

```
unit counter_checker (logic clk,
                      logic [7:0] counter);
  clock negedge (clk) {
    event overflow_ev :
       (cnt == 8'hff) #1 (cnt == 8'h00);
  }
  assert overflow23 : forbid(overflow_ev);
endunit

bind module counter_13bit :
                counter_checker (clk, cnt);
```

The assertion is implemented in the unit counter_checker using the OVA assert statement. The assert states that event overflow_ev may not occur. overflow_ev will occur if the value of cnt transitions from 255 on one clock negative edge to zero on the next. The assertion is bound to the Verilog module using the bind module statement.

The Property Specification Language was originally developed by IBM under the name Sugar and later donated to Accellera. Not unlike the OVA levels, PSL is partitioned into four layers: Boolean, temporal, verification and modeling. The Boolean layer is used to construct the expressions used by the other three layers. The name is somewhat of a misnomer because it defines many other types of expressions. Nonetheless, these expressions form the foundation of the temporal layer. The temporal layer is the core of PSL and is used to define properties. These properties may be either Boolean or temporal. The verification layer is used to specify the application of the properties defined in the temporal layer. For example, a property may designated either a checking property or a coverage property. The modeling layer is used to represent the behavior of device inputs for static analysis.

This is an example of a PSL checker assertion that mandates every request (`req`) is followed by an acknowledge (`ack`) on the rising edge of the next clock (`clk1`):

```
assert always (req -> next ack) @(posedge clk1);
```

Similarly, the following is a PSL coverage assertion that records each time the sequence *request→acknowledge→not request→not acknowledge* is observed on consecutive clocks:

```
cover {req; ack; !req; !ack} @(posedge clk1);
```

SystemVerilog assertions are a composite of OVA and PSL, supporting both procedural (immediate) and concurrent assertions. The following SystemVerilog assertion states that acknowledge may not be asserted on two consecutive clocks:

```
property not_2_acks;
  @(negedge clk)
    disable iff(reset) not(ack [*2]);
endproperty

assert property(not_2_acks);
```

Having surveyed the common kinds of assertions, what assertion coverage must be measured? First, the coverage of all checking assertions must be measured because each is responsible for detecting a property violation. The property may be a specification requirement or a designer's intent. Unless we know the assertion was executed or evaluated, we cannot be certain the device conforms to the property.

In addition, we must measure coverage of each coverage assertion that implements RTL-level functional coverage. This RTL-level functional coverage is a verification plan requirement. Unless we observe that a coverage assertion has been executed, we cannot distinguish missing device behavior from an unevaluated assertion when a coverage hole is reported.

Concluding this brief introduction to assertions and when their coverage must be measured, let's move on to the subject of this chapter: assertion coverage.

6.2. Measuring Assertion Coverage

The term "assertion coverage" has several meanings in industry today. Sometimes it is used to refer to the ratio of assertions to RTL lines within the design. I prefer to describe this as "assertion density" because it refers to the frequency distribution of assertions throughout the RTL. At other times, it is used to refer to functional coverage implemented using assertions. However, it should then be described simply as "functional coverage," implemented using coverage assertions.

In this book, "assertion coverage" means recording the fraction of assertions executed (or evaluated), passed and failed. First, I explain the life cycle of an assertion and then examine several assertion coverage implementations.

If an assertion is simulated (versus formally proved), it may be in one of two states, idle or evaluating, as illustrated below:

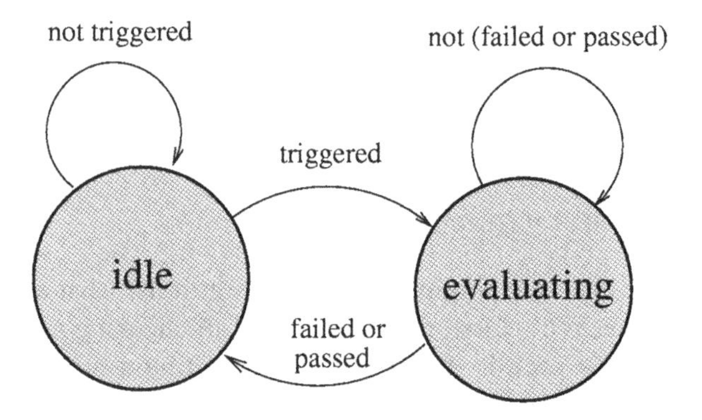

Figure 6-2 Assertion Finite State Machine

When idle, the assertion is awaiting its triggering event or expression. Once triggered, it moves to the evaluating state. If the assertion is a simple Boolean expression such as *not (request and grant)*, the assertion is triggered with each simulator event evaluation cycle and it immediately either succeeds or fails. If the assertion is an implication like *request implies grant on the next cycle* or a sequential expression[7] such as *request, then acknowledge, then grant*, the assertion is triggered when the antecedent (*request*) evaluates to true.

With this explanation of the assertion FSM, how is the behavior of an assertion recorded? The procedure for measuring assertion coverage varies for each assertion language and from one simulator to the next. Let's look at one in particular: OVL with Verisity Design's Specman Elite®.

6.3. Open Verification Library Coverage

Specman includes a coverage and assertion interface (CAI) that allows external coverage data to be displayed and analyzed — along with functional coverage — from the Specman coverage browser. Analyzing OVL assertion coverage from Specman is a three-step process.

The first step is instrumenting the Open Verification Library. The OVL is instrumented by the Verisity code coverage tool SureCov to enable recording the assertions that execute and those that fail during simulation. The instrumentation is only performed one time. Thereafter, the instrumented library is used with each design instead of the original library. (A pre-instrumented copy of the OVL is distributed with the CAI.) When an assertion fails during simulation, the DUV error management mechanism of ***e*** is notified so that the verification environment may handle the failure the same as any other device failure.

The second step is extracting OVL assertion instances from the RTL. An extraction program reads the design database and writes an interface file that is used by CAI to record per-instance assertion data. The extraction program is run once the RTL is stable and any time thereafter a new assertion instance is added to the RTL.

The third step is running the environment with the instrumented OVL. At the end of each simulation, the Specman coverage database (*basename*.ecov) will contain a record of the assertion coverage. The coverage results may be analyzed using the Specman coverage browser.

Having examined simulation-based assertion coverage measurement,

[7] PSL and SystemVerilog make use of Sugar extended regular expressions (SERE) to describe sequential Boolean expressions.

what is its static verification counterpart?

6.4. Static Assertion Coverage

In contrast to a simulated assertion, if an assertion is formally proved, the proof engine of the formal tool will either report it as proven or provide a counter-example that demonstrates the assertion is not satisfied. The counter-example is a sequence of inputs — typically from the device reset state — that lead to a violation of the assertion. These inputs may be used to derive the necessary vectors to simulate the failing scenario. Although unnecessary for validating the assertion violation, analysis of the sequence of device states that preceded the violation will shed light on the logic flaw or absence of logic responsible for it.

What if an assertion is not proven but a substantial fraction of its associated functionality has been examined by the proof engine? Is there a way to determine the coverage offered by the "partial proof" without resorting to decomposition of the assertion into more limited assertions? Many temporal assertions are synthesized into finite state machines for either static or dynamic evaluation. If the proof engine of a static tool calculates the states visited and arcs traversed while attempting to prove an assertion, that fraction of the total number of states and arcs may be used to infer partial assertion coverage.

Coverage of fully proven assertions is no more than a matter of maintaining a database of assertions proven and not proven, relative to the current RTL snapshot. With each new RTL release, previously proven assertions are invalidated so the proofs must be rerun.

6.5. Analyzing Assertion Coverage

The analysis of assertion coverage is dictated by whether or not an assertion is a checker assertion or coverage assertion. The checker assertion is inserted to detect the violation of a property or to prove the property is never violated. The coverage assertion is placed in the RTL to report the occurrence of an expected event. How do we analyze the record of their activations?

6.5.1. Checker Assertions

Checker assertions are responsible for detecting violations of data and temporal properties. These properties are specified for structural entities, interfaces and protocols. Each kind of property is illustrated in 6-3 below:

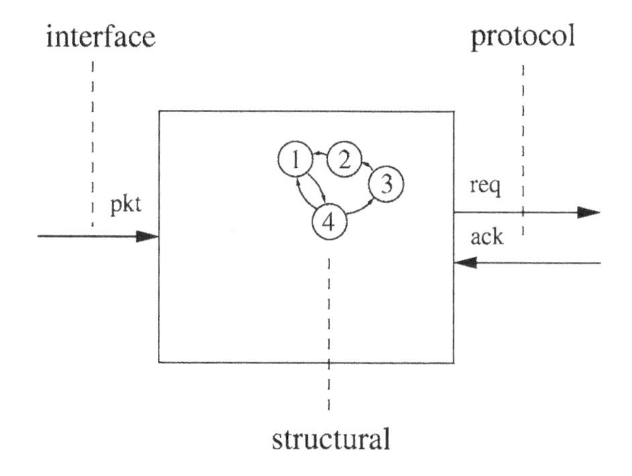

Figure 6-3 Structural, Interface and Protocol Properties

A structural entity is a building block such as an FSM or FIFO. An interface is a point at which independent blocks interact through a defined set of signals. A protocol is a set of rules governing communication between two or more blocks. Properties are defined for each of these kinds of properties.

Suppose after running all of your regressions, a structural assertion such as "FIFO pkt_fifo must never overflow" has never been executed. This is likely a concurrent assertion because it should be monitoring the FIFO state at all times. A concurrent assertion is activated by an associated event. If the event never occurs, the assertion will never be executed. If the event is a clock edge or level, something is mistakenly gating the clock to a static value. If the condition is checked less frequently, for example after each FIFO write, why was the FIFO never written? This could be an input coverage hole, indicating the stimulus generator is unable to cause the FIFO to be written. It could also be a logic bug where the device should have written the FIFO but did not. Each possibility must be considered before it is ruled out.

Now, let's look at an interface assertion. On the input interface of the block illustrated above (figure 6-3), the assertion "*pkt* must remain stable for two clocks" is defined. We have been running regressions for several days and have reached our functional and code coverage goals. All but one of our assertions have been observed: this particular *pkt* assertion. It is a procedural assertion, executed by the packet bus unit while processing incoming packets. Why hasn't it been executed?

There is a preponderance of evidence — so to speak — that packets *are* flowing through the device so the bus unit must be receiving packets. Examining the packet processing code, we discover there are now two control flow paths even though there was only one when the assertion was inserted. This precludes the assertion from executing except when it is a type-B packet. (Type-B packets were a late addition to the design.)

The solution is to move the assertion out of the conditional control flow path so that it is unconditionally executed for each incoming packet, independent of its packet type. The lesson to be drawn from this coverage hole is to place a procedural assertion in a control path so that it is tightly coupled to the RTL. **BIG BOLD LETTERS** letters should make it clear that the assertion is intertwined with the RTL code. Any changes to the assertion or the RTL necessitate careful examination of its associated pair.

What about analysis of protocol assertions? Suppose we have an assertion on the interface labeled "protocol" in figure 6-3 above. The assertion states "after *req* is asserted, *ack* must be asserted within 1 to 4 cycles." This is a concurrent checker assertion whose coverage indicates it has been evaluated many times. In fact, the coverage report indicates it has been evaluated an order of magnitude more times than the total number of cycles in all of the regressions.

A quick check confirms that *req*, the event that activates the assertion, is supposed to be synchronous with the system clock. Since it is activating this assertion more frequently than the clock frequency, *req* must glitching. Sure enough, it is. Sometimes unexpected coverage results reveal device errors. Review those results carefully!

6.5.2. Coverage Assertions

Coverage assertions are the implementation layer of a functional coverage model.[8] The model would be implemented using an assertion because it concerns itself with the RTL abstraction level. For example,the semantic description[9] of a coverage model may be "record all 3-packet permutations of type-A and type-B packets processed by the bus unit." After completing the model design process, the resultant English description of the coverage assertion might read "a three packet sequence entering the bus unit may be composed of any combination of type-A and type-B packets."

If this coverage assertion has not been executed (procedural assertion) or evaluated (concurrent assertion) during the regression runs, the same analysis required of a checker assertion must be applied. If this is a procedural assertion, the control flow path on which the assertion resides was never traversed because of a logic error, assertion error or missing stimulus. If it is a concurrent assertion, the activation event never occurred.

6.6. Summary

In this chapter I introduced assertions and assertion coverage. I discussed several ways of classifying assertions, the kinds of properties that may be specified and the various property specification languages. The different meanings of "assertion coverage" were distinguished from one another. Finally, you learned how to measure assertion coverage and analyze the results.

[8] The three layers, or steps, of functional coverage model development are top-level design, detailed design and implementation. See chapter 4, "Functional Coverage," for an in-depth discussion of the functional coverage model development process.

[9] See section 4.3, "Top-Level Design," for more information on the semantic description of a coverage model.

7. Coverage-Driven Verification

The preceding chapters laid the foundation for the methodology discussed in this chapter, coverage-driven verification (CDV). In this chapter, I introduce coverage-driven verification, address common objections to its use and then discuss stimulus generation, response checking, coverage measurement and coverage analysis in the context of CDV.

Coverage-driven verification is a verification methodology in which coverage planning[1] precedes the rest of the verification process. Coverage planning means defining a strategy for measuring verification progress — employing functional, code and assertion coverage — and the tactics that will be employed to implement it. All of the functional coverage models necessary to describe the device behavioral requirements are designed at the top-level. (The detailed design and implementation of some models may be postponed until more fundamental coverage goals have first been achieved.) The code coverage metrics to be recorded and analyzed are specified, along with code coverage goals aligned with design milestones. The assertion coverage strategy is devised, addressing both checker assertions and coverage assertions. Goals for full checker assertion coverage are specified for key design milestones. The partitioning of responsibility for implementing RTL level functional coverage between the HLVL and coverage assertions is specified. The full scope of the verification problem and the tool set to be employed by the verification team to measure each advance are specified *before* stimulus generation or checking are implemented.

Contrast this with directed testing, where an enumerated list of specific verification scenarios are first defined and then implemented as tests. Although both methodologies define corner cases of the device, CDV specifies these as broad goals and uses a generation module to stimulate the cases

[1] See chapter 2, "Functional Verification" and Verification Plans: The Five-Day Verification Strategy for Modern Hardware Verification Languages by Peet James, Kluwer Academic Publishers, 2003.

and those surrounding them. Directed testing chooses a particular instance of each corner case and derives an enumerated list of variants.

Before delving into the process of coverage-driven verification, let's address some common objections.

7.1. Objections to Coverage-Driven Verification

I know a lot of engineers and managers who object to using coverage-driven verification, for a variety of reasons. These include a belief their current methodology is adequate, skepticism about whether it really works, tools which don't support the process, no verification plans, immense verification spaces, not enough time for dealing with coverage and no value to coverage at the start of the project. Let me address each objection in turn.

"Our current methodology is good enough." This objection is generally premised on the belief that past success is a reliable indicator of future success. That is not necessarily true. Consider a new processor design that is a re-implementation of an existing instruction set architecture (ISA). The verification flow used to demonstrate ISA compatibility may be reused for the new processor because, by definition, the new processor must run the existing software base. However, a new micro-architecture is invented to meet more stringent performance and power requirements. This micro-architecture will come with a never-seen-before set of boundary conditions and bugs: a new verification problem. Unless the new verification problem is described, quantified and implemented using a coverage-driven approach, the probability of taping out the design with latent bugs is high.

"I don't think CDV really works." This objection, skepticism about the efficacy of coverage-driven verification, has its roots in two aspects of design culture: the need for measurable results and resistance to change. In my experience, management becomes uncomfortable when code is not being written or bugs are not being found. The desire to see measurable engineering results, such as verification environment code or RTL, rushes many engineering teams through design into implementation far too soon. This leads to poor implementation choices and unnecessary rework. Management needs to schedule time for verification environment design, which includes writing a verification plan and the associated coverage plan.

"There aren't any tools that support the CDV methodology." This is being addressed by EDA vendors by automating parts of the verification process. For example, in 2004 Verisity Design introduced *v*Manager, a tool for

managing the verification process through the use of an executable verification plan ("*v*Plan"). The *v*Plan is used to track coverage progress throughout the design process. Another EDA vendor, 0-In Design Automation, introduced the Archer verification system in 2004. This tool-set supports coverage-driven verification by integrating the measurement and analysis of assertion coverage and structural coverage. As this book goes to press, the SystemVerilog draft standard (3.1a draft 5) has introduced ***e***-like language constructs to implement functional coverage models. Commercial SystemVerilog simulators will soon support these constructs.

"We don't write verification plans so how can we employ coverage-driven verification?" The short answer is: You can't! The long answer is: It's time to start writing verification plans because, without a plan, you are skipping a vital step in the verification process. The verification plan outlines the scope of the verification problem and how the device will be verified. It also serves as the design specification for the verification environment. Reread section 2.4.1, "Functional Verification Plan," to make sure you understand the importance of the verification plan and what it should address.

"The size of our verification space is enormous!" Engineering teams today are building incredibly complex devices, such as single-chip cellular phones, MPEG players and PDAs. Hence, another common objection is that the combinations of functions to verify is endless! How do I know when I've written enough coverage? The verification space of modern chip designs is, indeed, enormous. However, by applying the techniques you've learned in this book you can pare this space down to a manageable size by structuring it into functional, code and assertion regions. Because the functional coverage is structured as organized attributes and the code and assertion coverage are automatically recorded by tools, the amount of manual effort required is minimized. Coverage hole analysis, discussed later in this chapter, is facilitated by advanced tools which address the process.

"We don't have enough time to verify the design, let alone measure and analyze coverage." This is like saying you don't have time to open a map and plan a route. You have places to go and people to see! Understanding the scope of the verification problem, designing a solution to solve it and measuring verification progress throughout the design cycle are as important, if not more so, than the other aspects of functional verification. Verifying a chip must start with defining what must be verified and planning how to accomplish it. The verification plan and resultant coverage models and goals serve as a road map for the verification of the design.

Another reason to lead with coverage is that designing functional coverage models up front helps to identify the requirements of the stimulus generator of the environment. Each input coverage model explicits describes the input scenarios and parameters that the generator must deliver. Another way of looking at this is that specifying requirements of the stimulus generator is akin to specifying input coverage models.

"There is no value in coverage planning at the start of the project." This engineer will worry about coverage when the project nears completion. This presents a catch-22 because, in order to *know* when the project is nearly complete, coverage is required. Yet, coverage will not even be considered until the project is almost done. If coverage is not planned before proceeding with the implementation of stimulus generation and checking, no "ruler" will be available to measure when "the project is almost done." Perhaps the start and end of a project are dictated by schedule alone — dates on a calendar — with no consideration for what fraction of the bugs have been found. This is a risky approach that will eventually lead to failure because the correctness of the design is not factored into the tape-out decision.

With the most common objections to adopting coverage-driven verification addressed, it's time to examine how stimulus generation, response checking and coverage measurement contribute to and are influenced by CDV. If you haven't read the earlier chapters[2] in the book on these subjects, I recommend doing so before proceeding.

7.2. Stimulus Generation

What role does stimulus generation play in a coverage-driven verification flow? As in other dynamic verification (simulation-based) methodologies, the primary purpose of stimulus is to exercise the DUV, causing it to exhibit behavior to be compared against reference behavior. The exhibited behavior is the result of the device visiting its operating states. However, in a CDV flow stimulus generation takes a back seat to coverage measurement because the defined coverage goals — functional, code and assertion — define the verification space to be visited. The responsibility of the stimulus generation aspect of the verification environment is to deliver full input coverage, internal coverage and output coverage. The input coverage is defined

[2] Stimulus generation is the subject of section 2.4.1.2; response checking is discussed in section 2.4.1.3 and coverage measurement is explained in section 2.4.1.1.

by the functional coverage models while code and assertion coverage measurement report whether or not the HDL is fully exercised.

Although stimulus generation is driven by defined coverage goals when using CDV, CDV does not preclude exploring device behavior outside of these defined goals. The remaining bugs in any design lurk in unseen places. These "places" are unforeseen or not-yet-explored operating conditions. The unforeseen scenarios may not be defined in coverage models so some random exploration of the device behavior is necessary.

I address two types of stimulus generation and their application to CDV in the following sections: conventional constraint-based generation and coverage-directed generation (CDG). Constraint-based generation defines a constraint set for calculating stimulus values and operates in an open loop, decoupled from coverage measurements. Coverage-directed generation may or may not define a constraint set but, more importantly, it operates in a closed loop, dynamically biasing its constraints based upon recorded coverage.

7.2.1. Generation Constraints

Since we are using an HLVL to implement our verification environment, generation constraints are employed to direct stimulus generation. The generation constraints are composed of two types: functional constraints and verification constraints. The distinction between them and how each is employed in a coverage-driven verification methodology is the subject the following two sections.

7.2.1.1. Functional Constraints

Functional constraints are those derived from the device functional and design specifications. The device functional specification defines its functional requirements. It answers questions such as:

- What capabilities must the device deliver?
- What is its feature set?
- What is its performance envelope?

The device design specification describes its architecture and high-level implementation. Questions such as:

- How many units compose the architecture and how are they organized?
- What communication protocol is employed between units?
- Is the design pipelined?
- If pipelined, how many pipe stages are there?

are addressed by the design specification.

In a coverage-driven verification flow, the device specifications are analyzed not only to understand the restrictions on generated input stimuli but also to determine the requirements for the input functional coverage models. These requirements determine attributes and their relationships.[3]

7.2.1.2. Verification Constraints

Verification constraints are those derived from the verification plan. They reduce the full space of valid stimuli to a subset useful for exposing device errors. This subset is characterized by device input boundary conditions and the data and temporal properties necessary to cause the device to activate internal boundary conditions and drive output boundary conditions.

Verification constraints in a CDV environment are initially specified in the verification plan but they are fluid throughout the design cycle. When a non-uniform distribution of values is selected to stress corner (or edge) conditions, it is usually based upon engineering judgement and intuition. As regressions proceed and the design is verified, coverage holes become apparent.[4] These holes expose coverage points or regions less likely to be visited than others. In order to balance the probability of visiting these points with those already observed, the generation weights need to be biased. In the absence of coverage-directed generation, the subject of the next section, biasing the generation weights is generally a manual process. However, it is possible to implement generation biasing feedback in a conventional HLVL environment.

In the following example, the 16-bit field `address` is periodically generated. The initial distribution of values is uniform because the generation weights (line 10) for the four ranges (0–15, 16–31, 32–47, 48–63) each have the value one (lines 11–13). During simulation, the list of weights is regenerated (lines 15–20) to adapt it to the currently recorded coverage.

---

[3] See "Top-Level Design" in chapter 4, "Functional Coverage."

[4] See section 7.5.3, "Hole Analysis."

```
1    type address_t : uint(bits:6);
2    type weight_t   : uint;

3    address : address_t;
4      keep soft address == select {
5         slot[0] : [ 0..15];
6         slot[1] : [16..31];
7         slot[2] : [32..47];
8         slot[3] : [48..63]
9      };

10   slot[4] : list of weight_t;
11     keep for each (weight) in slot do {
12       soft weight == 1
13     };

14   next_address() is {
15     gen slot keeping {
16       it[0] in ... ;
17       it[1] in ... ;
18       it[2] in ... ;
19       it[3] in ...
20     };
21     gen address keeping
22   }
```

The slot elements (generation weights) are biased during simulation to skew the probability distribution of addresses to those necessary to fill the remaining coverage holes.

7.2.2. Coverage-Directed Generation

An interesting duality and, I would argue, redundancy exists in a state-of-the-art constrained random verification environment. With respect to functional input coverage, generation constraints direct the stimulus aspect of the verification environment to that subset which is both functionally valid and useful for exposing device errors. At the same time, an input functional coverage model records what stimuli have been applied to the device, relative to the full set of desired stimuli. Specifying both the constraints and the input coverage is, in some sense, redundant.

For example, in *e* the constraint:

```
keep address == select {
    1 : [ 0..15];
    1 : [16..31]
}
```

restricts the values generated for field `address` to be within the range zero to 31. Half of the time, the generated value will be between zero and 15 and the other half of the time it will be between 16 and 31. The corresponding input coverage item:

```
item address using ranges = {
  range([ 0..15]);
  range([16..31])
}
```

specifies two buckets for the field `address`: one in the range zero to 15 and the other in the range 16 to 31.

In order to eliminate redundant specification of input requirements such as this and optimize the rate of functional coverage closure, coverage-directed generation infers the necessary generation constraints from the coverage specification and the remaining holes. At the time this book was written, coverage-directed generation is used internally by some companies but is not a commercially available technology. Let's look at an example drawn from my neck of the woods.

In the prairie land of north Texas, it is a challenge to grow healthy trees in the yard so we monitor rainfall and the soil characteristics. If I were using *e* to generate tree groves with proper conditions and I wanted to use coverage-directed generation, I might write a program like this.

First, I need some land for the groves (unit `sys`, field `grove`) and I need to specify how many groves I want to generate (`number_of_groves`):

```
extend sys {
 !grove              : grove_s;
  number_of_groves : uint;
    keep soft number_of_groves in [14..31];

  run() is also {
    for i from 1 to number_of_groves do {
      gen grove keeping { .grove_id == i }
    }
  }
}
```

This program will generate between 14 and 31 groves when it starts executing the "for" loop. The `number_of_groves` constraint is specified as `soft` so that it may be overridden from another constraint file. Each grove has a unique identifier (`grove_id`) constrained to its position in the sequence of generated groves.

What does a grove of trees look like in ***e***?

```
struct grove_s {
  number_of_trees         : uint;
  tree[number_of_trees]   : list of tree_t;
  soil_pH                 : uint[0..14];
  soil_consistency        : soil_consistency_t;
  moisture                : moisture_t;
  grove_id                : uint
}
```

It is a struct containing six fields:

`number_of_trees`	the number of trees in the grove
`tree[]`	a list of tree types
`soil_pH`	the acidity (or alkalinity) of the soil
`soil_consistency`	the blend of clay and sand in the soil
`grove_id`	a unique identify for the grove

The tree, soil consistency and moisture types are these enumerated types:

```
type tree_t                : [BOISDARC, HACKBERRY,
                              PEACH, PECAN, WALNUT];
type soil_consistency_t : [CLAY, SILT, LOAM,
                              SANDY];
type moisture_t            : [BONE_DRY, DRY, DAMP,
                              WET, SOAKED]
```

The number of trees generated in each grove is specified by:

```
extend grove_s {
  keep soft number_of_trees in [0..513]
}
```

The statement `extend grove` adds struct members to the existing struct `grove_s`.[5] In this case, I added a constraint to the generated value of `number_of_trees`.

Note that we have not yet specified any rules for generating the groves, trees and soil. The beauty of coverage-directed generation is that we don't have to. Instead, only the input coverage goals need to be specified:

```
extend grove_s {
  cover grove_created is {
    item soil_pH using ranges = {
      range([ 0.. 2], "", UNDEF,  5);
      range([ 3.. 5], "", UNDEF, 10);
      range([ 6.. 8], "", UNDEF, 15);
      range([ 9..11], "", UNDEF, 30);
      range([12..14], "", UNDEF, 40)
    };
    item soil_consistency using ranges = {
      range([ CLAY], "", UNDEF, 1);
      range([ SILT], "", UNDEF, 2);
      range([ LOAM], "", UNDEF, 6);
      range([SANDY], "", UNDEF, 1)
    };
```

[5] The ***e*** struct members are field, constraint, method, event, when, coverage group, on block and expect.

```
    item moisture using ranges = {
      range([BONE_DRY], "", UNDEF, 2);
      range([DRY      ], "", UNDEF, 5);
      range([DAMP     ], "", UNDEF, 20);
      range([WET      ], "", UNDEF, 5);
      range([SOAKED   ], "", UNDEF, 2)
    };
    item number_of_trees using ranges = {
      range([0],          "", UNDEF,  5);
      range([1],          "", UNDEF,  5);
      range([2..49],      "", UNDEF, 10);
      range([50..99],     "", UNDEF, 50);
      range([100..499], "", UNDEF, 15);
      range([500],        "", UNDEF,  3);
    };
    cross soil_pH, soil_consistency, moisture
  };

  event grove_created;

  post_generate() is {
    emit grove_created
  }
} // extend grove_s //
```

The coverage group `grove_created` is added to the struct `grove_s`. It defines four simple items — `soil_pH`, `soil_consistency`, `moisture` and `number_of_trees` — and one cross item: `soil_pH` × `soil_consistency` × `moisture`. These items are sampled each time the event `grove_created` is emitted. `grove_created` is emitted by the method `post_generate()` after a `grove_s` is generated.

Each simple item defines a set of buckets and each bucket has a specified range of values. The fourth parameter of the `range()` option, the *at-least* value, specifies the minimum number of times an item value in this range must be sampled before the coverage hole is filled. (The second and third parameters are unused options.) The relative *at-least* values imply a distribution of values for the item.

The result of using coverage-directed generation on generated input values is that the necessary input constraints are inferred rather than specified by the engineer. In order to apply CDG to other kinds of coverage — such as internal and output functional coverage, code coverage and assertion coverage — correlations must be discovered between applied input stimuli and coverage holes. At the time this book was written, Bayesian networks are being successfully applied to construct these correlations.[6]

Having explored the CDV impact on stimulus generation, how is the checking aspect of the verification environment influenced?

7.3. Response Checking

What demands are placed upon response checking in a coverage-driven verification environment? Unlike in a directed test environment, in a CDV environment data and temporal checking are generalized and broad in scope. In addition, there is a trade-off between the fidelity of checking and the fidelity of coverage. Each is discussed in turn.

When using a directed test methodology, each test is usually responsible for applying stimulus and checking the response of the DUT to the applied stimulus. With knowledge of the specific applied stimulus, the test author is able to take shortcuts in the checking section of a test.

For example, years ago my verification team designed tests with expected results and observed results data sections. The expected results section was a sequence of value sets, one per test case. When the test ran, each of its test cases wrote the device response values into the observed results section. When the test finished, a component of the runtime environment compared the expected results to observed results sections to determine whether or not the test passed.

In a coverage-driven verification environment, the checking aspect runtime behavior is completely decoupled from stimulus generation. As such, it must check all of the functional requirements demanded by the device specification. Both data and temporal checks, on all interfaces, are required.

---

[6] The reader is encouraged to review the leading-edge work of IBM Haifa Research Labs (http://www.haifa.il.ibm.com/) on coverage-directed generation. The paper "Coverage Directed Test Generation for Functional Verification Using Bayesian Networks," by Shai Fine and Avi Ziv, was published in the proceedings of the 40th Design Automation Conference, June 2003.

These functional requirements are derived from the same attributes and attribute relationships as the functional coverage models. As a coverage model is designed — and recorded in the verification plan — the device behavioral requirements should also be recorded.

The detailed design and implementation of the checking aspect should also leverage the data and temporal monitors designed for the coverage models. In figure 7-1 below, the checker and coverage modules share the input and output monitors.

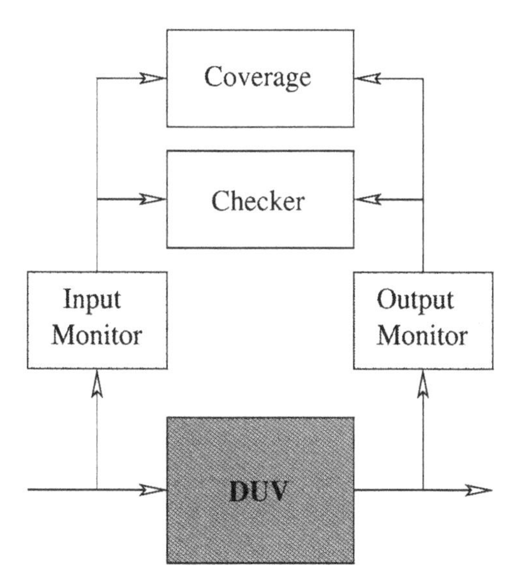

Figure 7-1 Checking and Coverage Aspects

Each obtains its data and temporal information from the monitors.

Another consideration for checking in a CDV environment is the tradeoff between the fidelity of the checking and coverage components. By "fidelity," I am referring to how precisely a component examines or records its values. For example, the checking module might only examine the primary outputs of a block[7] or it could also examine internal nodes. The latter case would be a higher fidelity design than the former.

[7] I use "block" in this book to refer to a section of RTL corresponding to a single functional unit designed by one engineer, as illustrated in a block diagram of a device functional specification. The term "functional unit" is usually synonymous with "block."

The trade-off between checking and coverage fidelity is a function of the observability into the device. If we are using pure black box checking but want to ensure a particular internal corner case was exercised, white box coverage is required to confirm the corner case was exercised and its result propagated to a primary output. On the other hand, if white box checking is employed, white box coverage must only confirm the corner case was exercised, not whether is was propagated to a black box checker.

7.4. Coverage Measurement

Coverage measurement and its complement, coverage analysis,[8] are the cornerstones of coverage-driven verification. How do we employ the various kinds of coverage in a CDV environment? In the preceding chapters I discussed functional coverage, code coverage and assertion coverage in depth.[9] In the following sections, I explain how this coverage should be used within the context of a CDV methodology.

To establish the context for the application of these coverage techniques, let's first take a look at the rate at which the RTL is written. The amount of code written versus time has a curve something like this (figure 7-2:

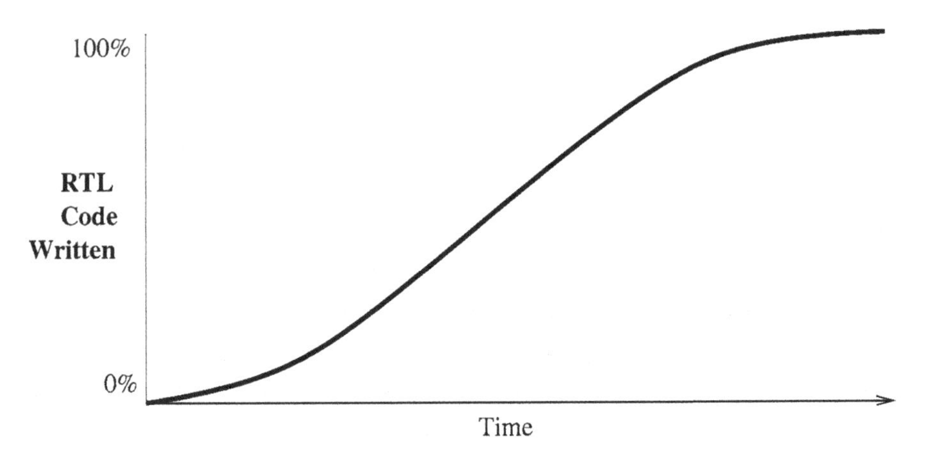

Figure 7-2 RTL Coding Progress

[8] See section 7.5.

[9] Functional coverage measurement is discussed in chapter 4; code coverage in chapter 5 and assertion coverage in chapter 6.

RTL development starts off slowly as the design specifications are still in flux, ramps up to a near-linear rate and then tapers off as the implementation nears completion.

The corresponding change rate of RTL over time is the slope of the previous curve (figure 7-3):

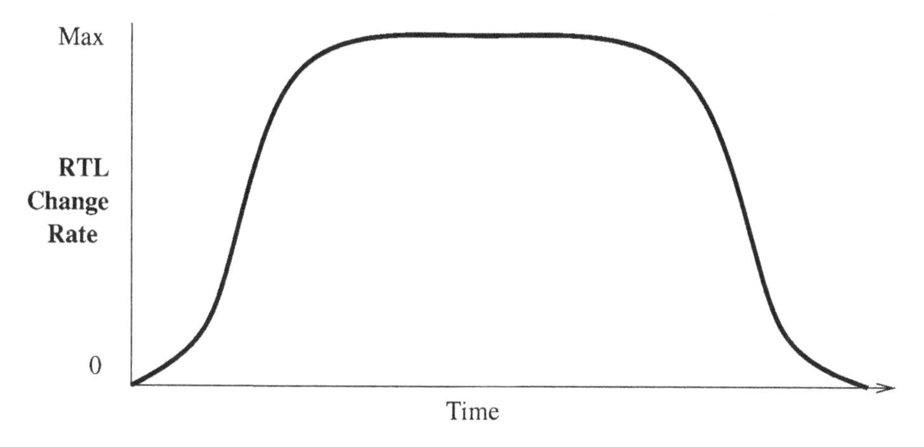

Figure 7-3 RTL Change Rate

The volatility of the RTL is a factor in choosing when to employ each of the coverage types. Now let's look at each in turn.

7.4.1. Functional Coverage

Functional coverage is used to explicitly specify specification and implementation metrics to be recorded. In a CDV methodology, functional coverage drives the rest of the verification process and, as such, must lead the other activities. Recall from chapters 2 and 4 that functional coverage models are the product of writing a verification plan. Each model quantifies part of the device verification space.

Figure 7-4 below illustrates the rate at which functional coverage should be developed over time:

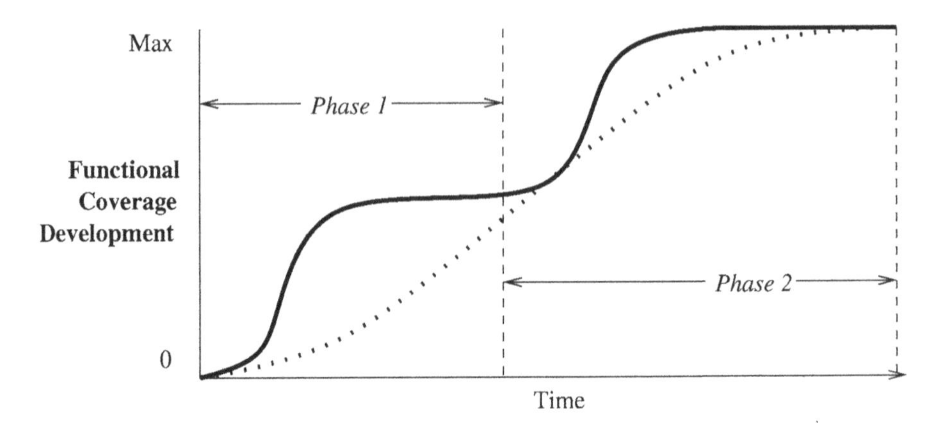

Figure 7-4 Functional Coverage Development Progress

The solid line is implemented functional coverage versus time. For reference purposes, the suppressed dotted curve is the amount of RTL code written over time. (This same curve also appears on the following two figures.) The verification plan defines a set of coverage models of a certain fidelity. These models should be implemented during the first phase of the design cycle, labeled *Phase 1* above. Once these models are implemented and you are approaching 100% coverage, you should begin refining the models to improve their fidelity as illustrated in *Phase 2*. The fidelity of a coverage model is improved by adding interacting attributes and increasing the number of ranges and the size of the ranges of attributes with value ranges. This will necessarily increase the size of the coverage spaces (i.e. number of coverage points). However, your autonomous verification environment, deployed across a large regression ranch, will quickly traverse this new space. New coverage models may also be designed and implemented in order to capture newly discovered attribute relationships.

This cycle of refining your coverage models, adding new models and rerunning regressions until you reach 100% coverage of the new models should be repeated until the fidelity of the coverage models is satisfactory.

7.4.2. Code Coverage

Code coverage records implicit implementation metrics of the RTL.[10]

[10] Code coverage measurement is discussed in chapter 5.

In CDV environment, aside from occasional use by designers while they are writing their RTL blocks and running localized sanity checks, code coverage should not be widely employed until the RTL begins to stabilize. This is illustrated in figure 7-5 below.

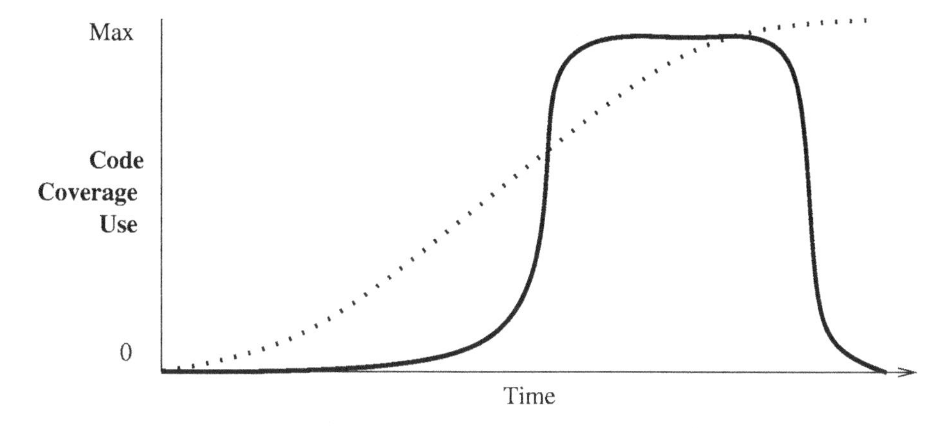

Figure 7-5 Code Coverage Use

Your objective in measuring code coverage is to discover implemented functionality which has not yet been exercised by the verification environment. While the RTL is changing, attempting to reach 100% code coverage is aiming at a moving target. Wait until the design team is finished implementing before assessing how well your verification environment exercising the RTL.

Some functionality you might measure in a functional coverage model may already be recorded by a code coverage metric. For instance, the behavior of a finite state machine often implements a function of the device. Rather than write a functional coverage model for the FSM to record state visitation and arc traversal, the FSM metric of a code coverage tool should be employed.

Code coverage also provides a cross-check against functional coverage models. Once the RTL has stabilized, code coverage of RTL implementing specific functional requirements should be compared against the corresponding functional coverage. Discrepancies may reveal flaws in a functional coverage model (code coverage but hole in corresponding functional coverage) or superfluous RTL (code coverage hole but corresponding functional coverage).

Contrasting the application of code coverage and functional coverage, it should be apparent that the effort required to use code coverage is back-end loaded while functional coverage is front-end loaded. Depending upon the number of metrics enabled, a code coverage tool may report an enormous amount of data that, through analysis, becomes information to be digested and filtered. On the other hand, no effort is required to start using code coverage. Functional coverage use is front-end loaded because the coverage models must be designed and implemented before any measurements are available. However, the reported results require little, if any, filtering because each model only records necessary functional observations.

7.4.3. Assertion Coverage

Assertion coverage, which records which assertions were executed, passed and failed, provides detailed insight into how well low-level design intent was preserved in the RTL.[11] In a coverage-driven verification flow, assertion coverage measurement necessarily lags RTL development because the assertions are written by the design team, into the RTL. This is illustrated in figure 7-6 below.

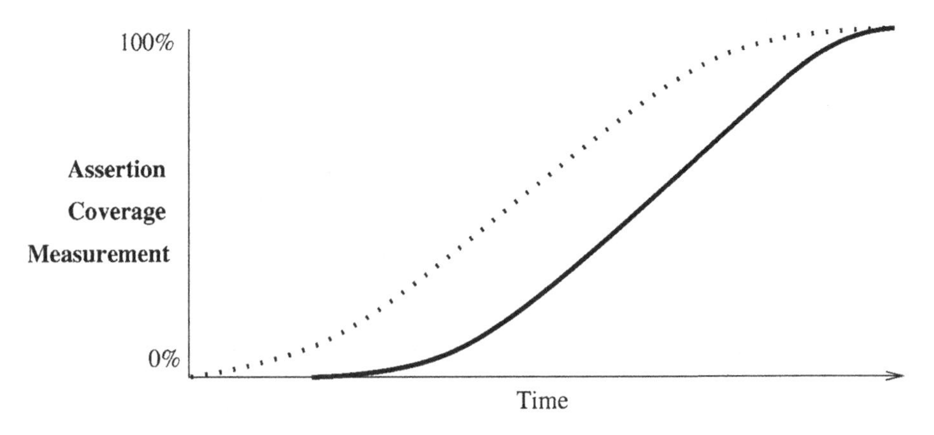

Figure 7-6 Assertion Coverage Measurement

Since assertions check or record behavior at a higher abstraction level than the RTL implementation itself, they should be written before the associated RTL. However, assertion coverage should be measured after the associated

[11] Assertion coverage measurement is discussed in chapter 6.

RTL is written.

7.4.4. Maximizing Verification Efficiency

In order to reach coverage closure — full functional, code and assertion coverage — as soon as possible for each stage of the design cycle, the rate at which coverage metrics are observed must be maximized. In this section, I explain verification efficiency and how to maximize it using coverage-driven verification.

Verification efficiency can be defined in a number of different ways. Since an efficiency is a ratio — numerator over denominator — let's look at some candidate terms. Some potential numerators are:

- lines executed
- statements executed
- branches exercised
- conditional expressions evaluated
- events emitted
- assertions evaluated
- assertions passed
- FSM states visited
- FSM arcs traversed
- functional coverage points visited
- number of bugs found
- number of bugs fixed

Each of these are measures of verification progress, assuming the requisite stimulus generation and response checking are present.

The code coverage metrics *lines executed*, *statements executed*, *branches exercised*, *conditional expressions evaluated* and *events emitted* all provide visibility at the RTL implementation level. *Assertions evaluated* and *assertions passed* each provide insight into how well design intent, captured by the design team, has been exercised. Likewise, the FSM metrics *states visited* and *arcs traversed* allow us to see how well the device was exercised, as seen as a set of concurrent finite state machines. *Functional coverage points visited* is a direct measure of the amount of device functionality verified. The bug-tracking metric *number of bugs found* is a clear indication of RTL quality improvement. However, it does not tell us how many bugs remain, a crucial indicator of verification progress. Likewise, *number of bugs fixed* is an important metric to balance against bug discovery rate but it also does not hint at remaining bugs.

Since our aim is to maximize the rate at which we achieve functional closure, candidate denominators of the efficiency equation are typically a time measure such as *simulation cycles* or *wall clock time*. Between the two, simulation cycles has the advantage as a denominator because the performance of a number of factors — such as simulation speed and simulation platform speed — is removed. The verification engineer has little control over simulation speed because it is largely determined by the performance of the simulator and the hardware platform on which it runs. For this reason, we will discard wall clock time and adopt simulation cycles.

This leaves the numerator terms above, per simulation cycle, as verification efficiency measures:

- lines executed per simulation cycle
- statements executed per simulation cycle
- branches exercised per simulation cycle
- conditional expressions evaluated per simulation cycle
- events emitted per simulation cycle
- assertions evaluated per simulation cycle
- assertions passed per simulation cycle
- FSM states visited per simulation cycle
- FSM arcs traversed per simulation cycle
- functional coverage points visited per simulation cycle
- number of bugs found per simulation cycle

In order to maximize the rate of coverage closure, we need to increase the value of each of the efficiency measures. Other than "number of bugs found per simulation cycle," each of the metrics above is referred to as a coverage density. As defined in chapter 1, "The Language of Coverage," coverage density is *the number of coverage metrics observed per simulation cycle.* Increasing the value of an efficiency measure is equivalent to increasing its coverage density.

A maximum coverage density would be one in which a unique value of every independent coverage metric is observed on each simulation cycle. For example, a different RTL line would be executed each cycle and the same line would not be executed twice for the same device state. A different branch path would be executed each cycle. A unique FSM state in each independent state machine would be visited, and so on. Although such a maximal coverage density is not achievable in practice, we can approach it through coverage analysis and stimulus feedback. The following sections delve into the analyses and techniques required to maximize coverage density.

7.5. Coverage Analysis

As you measure verification progress using various coverage metrics, the measurements must be interpreted. The dissecting, review and interpretation of coverage measurements is coverage analysis. In this section you will learn how to analyze and interpret coverage measurements and make changes to the generation and coverage aspects of your verification environment to maximize verification efficiency and ensure coverage closure.

As mentioned above, verification efficiency may be maximized by minimizing observations of redundant metric values. We can approach the goal of observing unique metric values each simulation cycle by minimizing redundant logic execution. Applying coverage-directed generation[12] is the most effective way to minimize redundant execution because coverage measurements from the current simulation and past simulations are used to dynamically adapt generation constraints. In the absence of coverage-directed generation, an analogous process is employed manually. In the following section, I discuss automated and manual generation feedback. The following two sections address coverage model feedback and hole analysis.

7.5.1. Generation Feedback

Generation feedback refers to adapting the generation aspect of the verification environment so that the probability distribution of coverage metrics meets the requirements of the verification plan. For example, suppose the verification plan weighted all coverage points within the set of coverage models equally. The initial design and implementation of the stimulus generator would be unlikely to result in such a balanced probability distribution.

[12] See section 7.2.2, "Coverage-Directed Generation."

Likewise, perhaps the code coverage goals require the device arbitration module to be executed three times more frequently than the packet collision detector. Again, the stimulus generator is unlikely to meet this goal straight out of the chute. Assertion goals are hindered by the same process.

Generation constraints are biased to achieve these goals, either automatically or manually. If coverage-directed generation is employed, current and past coverage measurements are used by the constraint solver. The generation constraints are biased to raise the probability of observing the remaining coverage holes. In addition, some coverage goals may require changes or additions to the temporal generation aspect of the verification environment. In an ***e*** environment, this means sequences may need to be enhanced or added.

If coverage-directed generation is not used, functional, code and assertion coverage results are periodically reviewed. Input functional coverage holes are filled through manual biasing of generation constraints. Output and internal coverage holes need to be traced back to input scenario parameters. These parameters — data, temporal or both — should be biased to increase the probability of activating the device logic associated with these coverage holes.

7.5.2. Coverage Model Feedback

In addition to the generation feedback path discussed above, a second feedback path in the verification environment is from measured coverage back to the coverage models. While analyzing your coverage results, you may find that your coverage models need refinement. Some models were initially written to ascertain what regions of the complete verification space are being exercised. They were exploratory. Now, you are ready to choose some as necessary and discard the others as not relevant. What is the process for distinguishing between them and making the necessary changes?

As discussed earlier, coverage fidelity plays a large role in exploratory and production coverage models. An exploratory model is constructed to map the execution space of the device against its functional and design specifications. It is generally a coarse, low fidelity model, typically structured as a matrix model. It is not intended to be filled but only to serve as a grid against which a trace of device execution in its functional domain may be visualized.

A production coverage model, on the other hand, is intended to precisely quantify the functional requirements of the device and has specific coverage goals assigned to it. It may initially be a low fidelity model with a modest coverage goal of, say, 80%. Later during the design cycle, the coverage model is refined to more precisely model the input, output or internal behavior of the device. It will usually be a larger model, having more coverage points, and define a higher coverage goal, perhaps 95 to 100 percent. This process may be repeated a number of times at intermediate design cycle stages.

While analyzing the results recorded by an exploratory coverage model, you may discover device behavior mandated by its specification but not captured by a production coverage model. This behavior could either be rolled into an existing production model or the exploratory model could be modified into a production model. If an existing production model shares most of the attributes of the exploratory model, the missing attributes should be added to the model and its relationships extended to include the useful exploratory region. If no production model is related to the exploratory model, the exploratory model should be refined into a production model. This will usually entail transforming it from a matrix model to a hierarchical or hybrid model and pruning out the regions representing unnecessary behavior.

7.5.3. Hole Analysis

As you review your functional coverage measurements, you will find coverage holes: required stimuli or device behavior not yet observed. How do you go about figuring out what these holes have in common in order to plug them? In this section I present coverage hole analysis techniques first reported by the IBM Research Laboratory in Haifa, Israel at the 2002 Design Automation Conference[13] and explain how they may be applied to hierarchical and hybrid coverage models, as well as the matrix models discussed in the paper.

There are two kinds of coverage holes discovered while analyzing functional coverage: valid holes and invalid holes. A valid coverage hole represents an intended observation of an input, output or internal scenario that has not been observed. On the input, this could be due to a bug in the

[13] "Hole Analysis for Functional Coverage Data," by Oded Lachish, Eitan Marcus, Shmuel Ur and Avi Ziv, DAC 2002, http://www.research.ibm.com/-pics/verification/ps/holes_dac02.pdf (March 2004). See also "Defining Coverage Views to Improve Functional Coverage Analysis," Sigal Asaf, Eitan Marcus, Avi Ziv, DAC 2004.

stimulus generator. On the output or internally, it could be caused by a device error.

An invalid coverage hole is an unintended functional requirement in one of the coverage models. It is a bug in a coverage model, perhaps caused by a misunderstanding of the device specification by the coverage model designer, wherein an impossible behavior is expected to be seen. An invalid coverage hole is converted to a restriction on its model, either a restricted set of values of an attribute or a refinement of the relationship between one or more attributes.

Coverage hole analysis is greatly facilitated through coverage visualization. Consider the following user interface (figure 7-7) from a stock market mapping tool[14]

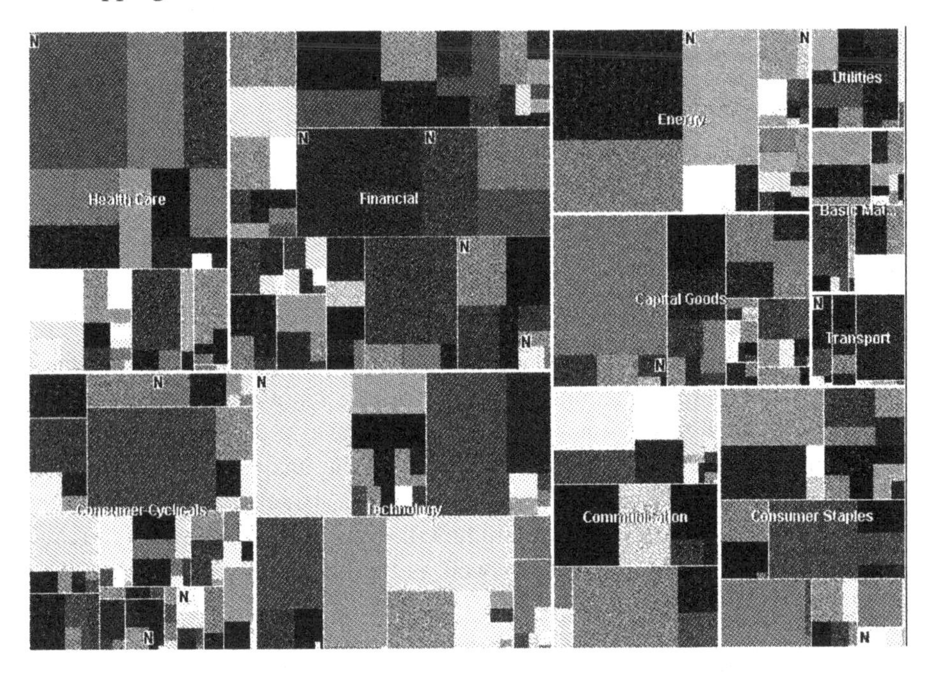

Figure 7-7 Overall Map of the Stock Market

Each of the major sectors, such as Financial and Technology, are labeled and have a white border. The area of the sector is proportional to the size of the market sector. The smaller rectangles within a sector represent companies

[14] http://www.smartmoney.com/marketmap/ (March 2004).

and the area of each is proportional to its market capitalization. The shade of gray in a rectangle indicates the performance of the stock of the associated company (black is down, white is up). (The actual user interface uses a red-to-green color scale where red means the stock price is down and green means it is up.)

If a mouse cursor is positioned over a rectangular region, additional information about the associated company is displayed, as illustrated in figure 7-8 below.

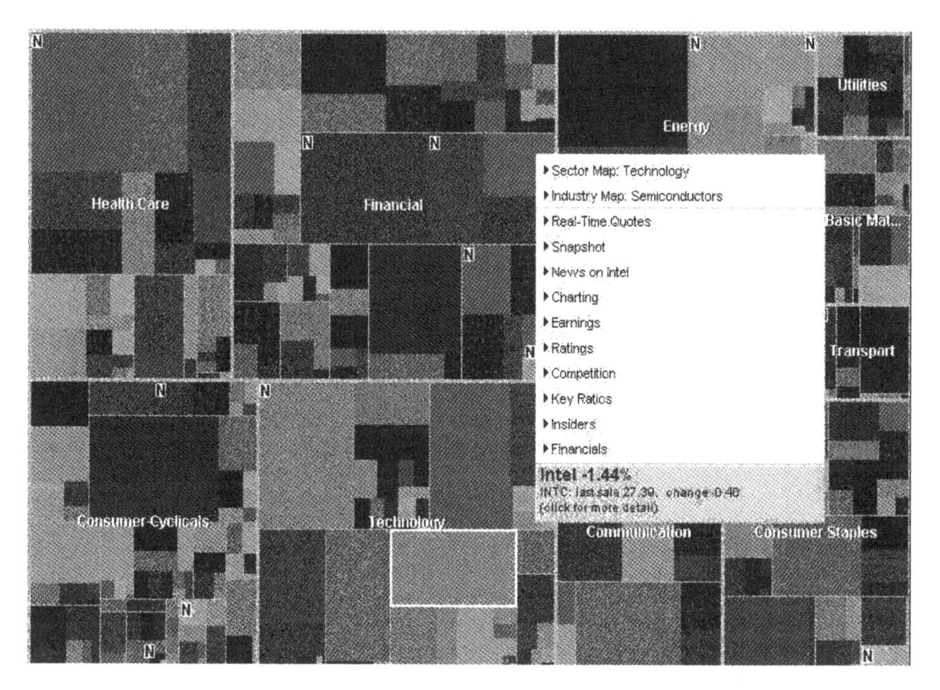

Figure 7-8 Stock Market Map Company Details

The text "Intel -1.44%, INTC: last sale ..." is displayed. If the text is selected with the mouse by clicking it, the illustrated pop-up menu is displayed, allowing the user to drill further down into the details of interest.

I use this as an excellent example of the use of visualization as a conduit to deep understanding and analysis of data.[15] Applied to coverage hole

[15] For the reader interested in further pursuing data visualization, I highly recommend these books by Edward R. Tufte: *The Visual Display of Quantitative Information*, 2001 and *Envisioning Information*, 1990, both published by

analysis, the area could represent an aggregate set of coverage models. The white border regions become coverage sectors such as instruction issue, operand types and bus interface. White regions would indicate full coverage while black would represent zero coverage. Further probing with the mouse reveals as much detail as necessary for the user to understand coverage holes and their locality.

In order to determine the commonalities among coverage holes and classify them, a useful metric is the proximity of one hole to another. Hamming distance is defined as the number of attribute values which differ between two coverage points. For example, consider two coverage points in the matrix model composed of attributes execution mode, opcode and register. If one hole has the value (real, ADD, AX) and the other (real, ADD, BX), the Hamming distance between the two holes is one because they differ in one attribute value: register. If one hole is (protected, JMP, CS) and the other is (protected, CALL, SS), the Hamming distance between them is two because they differ in two attribute values.

Coverage holes may be classified three ways: aggregation, partitioning and projection. Aggregation is the process of coalescing a number of individual coverage holes to create a smaller number of larger holes. In the following example, ten individual coverage holes are aggregated into one coverage hole region in two steps. Each coverage hole is defined by three attributes.

$$
\begin{array}{lll}
(2, 0, 1) & & \\
(2, 2, 1) & & \\
(2, 3, 1) & (2, \{0,2,3,9\}, 1) & \\
(2, 9, 1) \rightarrow & & \rightarrow (\{2,5\}, \{0,2,3,9\}, 1) \\
(5, 0, 1) & (5, \{0,2,3,9\}, 1) & \\
(5, 2, 1) & & \\
(5, 3, 1) & & \\
(5, 9, 1) & &
\end{array}
$$

Figure 7-9 Coverage Hole Aggregation

Partitioning is another way of classifying similar coverage holes. However, rather than grouping holes using a quantitative measure like attribute value, partitioning groups holes according to their semantic

Graphics Press.

similarity. In the same manner the top level design of a functional coverage model begins with its semantic description, the author of a model may also ascribe a meaning to attribute values or value ranges. This enables analysis of coverage holes to use partitioning to group together holes with common semantics.

The third way of classifying coverage holes is through the use of projection. When a coverage hole has one or more attributes, none of whose attribute values have ever been observed, it is referred to as a projected hole. The dimension of the projected hole is equal to the number of attributes whose values have never been observed. The term *dimension* is derived from observing that a matrix coverage model composed of N attributes is of order N. That is, it defines N dimensions. A coverage hole specified by N-1 fixed attribute values and one wildcard attribute (i.e. none of its values have been observed), defines a line in the N-dimension space. A hole with two wildcard attributes defines a plane, and so on. One goal of coverage hole analysis is to identify high-order projected holes and determine why their wildcard attributes have not been observed.

When coverage holes are determined to be erroneous holes, as described earlier, it should be easy to transform the hole specification into a restriction or constraint on the associated coverage model. For example, suppose I have the matrix coverage model EFLAGS.CF $\times$ EFLAGS.PF $\times$ EFLAGS.TF (carry flag, parity flag, trap flag). The model is designed and implemented as a cross coverage item in ***e***:

```
eflags_cf : bit;
eflags_pf : bit;
eflags_tf : bit;

cover eflags_written is {
  item eflags_cf;
  item eflags_pf;
  item eflags_tf;
  cross eflags_cf, eflags_pf, eflags_tf
}
```

Later, I discover the coverage hole (`eflags_cf`=0, `eflags_pf`=0, eflags_tf=0) is an erroneous hole. If my hole analysis tool identified the hole using the specification:

```
cover eflags_written is also {
  item cross__eflags_cf__eflags_pf__eflags_tf
    using also ignore = (eflags_cf == 0
                         and eflags_pf == 0
                         and eflags_tf == 0)
}
```

I could load the specification with the rest of my verification environment, turning the erroneous hole into an additional restriction on the coverage model. The item name cross__eflags_cf__eflags_pf__-eflags_tf is implied by the original cross item definition, "cross eflags_cf, eflags_pf, eflags_tf."

The techniques described in this section for grouping coverage holes and analyzing them may be applied to hierarchical and structural hybrid[16] coverage models as well. If analyzing a hierarchical model, each level of the hierarchy represents an attribute and each arc a value (or range of values) of the attribute. The same n-tuple specification used for the matrix model is employed for the hierarchical model. Likewise, because a structural hybrid model is composed of matrix and hierarchical regions, it may be represented and analyzed in the same way.

7.6. Summary

In this chapter I explained coverage-driven verification, the implications on stimulus generation, checking and coverage measurement and how to analyze coverage results. You learned that stimulus generation, with its functional and verification constraints, needs to meet the requirements of the coverage specification. It must also adapt to recorded coverage as the design cycle proceeds. Both conventional stimulus generation with manual coverage feedback as well as coverage-directed generation were explained.

The checking aspect of the verification environment is completely decoupled from the coverage aspect. This means that, although the two

[16] I qualify "hybrid" with "structural" to distinguish a model having a hybrid structure from a model composed of heterogeneous metric sources. The latter hybrid model is described in the next chapter, "Improving Coverage Fidelity With Hybrid Models."

aspects may share the same data sources (data and temporal monitors), response checking is solely responsible for the comparison process of dynamic verification. I also discussed the trade-off between coverage fidelity and checking fidelity.

The concept of verification efficiency was introduced. Various ways of maximizing verification efficiency were discussion. You learned that coverage density must be measured in order to maximize efficiency.

Lastly, I explained how to analyze coverage and, in particular, coverage holes. This analysis provides the necessary insight into adjusting the generation and coverage aspects of your verification environment in order to meet your verification goals.

8. Improving Coverage Fidelity With Hybrid Models

In this chapter I explain coverage fidelity, the motivation for employing high fidelity coverage models and propose one way of constructing them. At the time this book was written, no commercial tools support specification of the model described in this chapter. Therefore, I take the liberty to posit an extension to the ***e*** language as an implementation vehicle for illustrating such a coverage model.

The fidelity of a coverage model is a measure of how closely the model reflects device behavior. If it is an input functional coverage model, a high fidelity model captures all possible input scenarios within its space. A low fidelity model defines a small subset of the input scenarios or groups different scenarios into the same class, making them indistinguishable from one another in recorded results.

If the model is a high-fidelity output coverage model, all possible responses of the device to applied stimuli have corresponding points or regions within the model. A low fidelity model either defines a subset of device responses or groups larger classes of responses into common coverage points or regions.

Whether a coverage model has high fidelity or low fidelity, it usually captures a subset of the near-infinite permutations of values and sequences in stimuli and responses. However, the high fidelity model is much more likely to highlight a bug-exposing scenario than a low fidelity model. How may we use coverage fidelity to improve the verification process?

The functional errors in a design are inserted throughout the design cycle. At the same time, functional errors are discovered, diagnosed and removed (corrected). Some bugs have a large effective "footprint" in the coverage spaces and others do not. For example, if all virtual-8086 mode exceptions cause a verification environment checker to fire, indicating a device error, a coverage model recording exceptions and processor execution modes requires only one hit in virtual-8086 mode to expose the bug. This bug has a large footprint.

On the other hand, if a processor device only malfunctions when a CALL instruction takes a stack segment fault in protected mode, at the same time an external priority-32 interrupt is asserted, this represents a very small region in a coverage model. The bug footprint is very small.[1]

A low fidelity coverage model may not distinguish two related, but different, execution scenarios from one another. One may be processed properly while the other is not because of a functional bug. Within the other constraints the verification engineer operates within, the design and implementation of a high fidelity model is preferred over a low fidelity model. One solution to a high fidelity model is the hybrid model.[2]

In the next section, I walk through the design and implementation of a hybrid coverage model. In the following section, I discuss the overlap of functional, code and assertion coverage. Finally, in the last section I touch on coverage delivered by static verification methods.

8.1. Sample Hybrid Coverage Model

Up until this point in the book, each type of verification coverage has been presented as orthogonal to the others. However, recalling the discussion of the taxonomy of coverage in chapter 3, "Measuring Verification Coverage," what if we wanted to more precisely describe device behavior by composing heterogeneous coverage metrics? The metrics would be drawn from all four kinds of coverage spaces:

- Implicit implementation
- Implicit specification
- Explicit implementation
- Explicit specification

The result would be a hybrid coverage model, the subject of this chapter.

For an example, let's consider a coverage model whose requirement is to record all permutations of instructions fetched, execution paths through the instruction decoder, processor execution mode and instruction fetch buffer

[1] Perhaps as small as that of the six-legged variety!

[2] This hybrid model is distinguished from the structural hybrid model introduced in chapter 4, "Functional Coverage."

overflow assertion evaluated. This is the semantic description of the coverage model.
The top-level design of the coverage model requires the following attributes, attribute sources and sampling times:

Attribute	Source	Sampling Time
Fetched instruction	fetch bus monitor	instruction fetch
Instruction decoder paths	RTL source code	thread execution
Processor execution mode	RTL register	instruction decode
Instruction fetch buffer overflow assertion evaluated and passed	RTL assertion	evaluation

The fetched instruction may be one of: ADD, AND, CALL, IRET, JMP, LOAD, MOV, NOP, NOT, OR, RET, STORE, or SUB. It is captured from a monitor on the output of the instruction generator in the verification environment each time an instruction is fetched.

The instruction decoder paths are the control flow paths through the instruction decoder of the DUV. These are recorded by our code coverage program each time an RTL execution thread executes a path.

The processor execution mode may be real or protected. It is captured from a single-bit register in the device that is sampled by a monitor in the verification environment on the first cycle of instruction decode.

The instruction fetch buffer overflow assertion is an OVL assertion written by the designer. It is evaluated by the simulator and the number of times it is evaluated, passed and failed is recorded and reported to the verification environment.

Since the model definition requires a full permutation of the attributes, this model is structured as a matrix.

Now, we proceed to the detailed design of the model, in which its architecture is specified. Recall from chapter 4, the three questions "what?," "where?" and "when?" must be answered. What must be sampled for each attribute? Where should attribute data be sampled from? When should the attributes be correlated? The answers for this model are summarized in table

8-1 below.

Attribute	**What to Sample?**	**Where to Sample?**	**When to Correlate?**
instruction	Field `instruction` of the unpacked, decoded instruction	***e*** instruction fetch monitor	Whenever the instruction decode path is traversed, followed by the evaluation of the overflow assertion, followed by sampling the execution mode. The event `instruction_interaction_e` will be defined as this time.
decode path	Control flow path through instruction decoder	Verilog instruction decoder module	
exec mode	Processor execution mode	***e*** coverage unit	
overflow	Buffer overflow assertion executed and succeeded	Verilog module defining the assertion	

Table 8-1 Hybrid Coverage Model Detailed Design

The implementation strategy somewhat mirrors the detailed design. Each attribute will be sampled from its defined source at its own sampling event. Whenever the correlation event is emitted, the most recently sampled values of the attributes will be recorded in the coverage group.

The implementation of the model begins with extending an existing coverage unit to add the coverage group `instruction_interaction_e`.[3]

[3] The "_e" suffix is a coding style to indicate the name of an ***e*** event. Note that I've extended the ***e*** language in the following example in order to illustrate how a hybrid coverage model might be implemented in ***e***.

```
extend coverage_u {
  cover instruction_interaction_e is {
    ...
  }
}
```

A coverage group in *e* is identified by its sampling event: instruction_-interaction_e. (This is, however, the correlation event of the coverage model.) instruction_interaction_e is defined later. In this coverage group, we will define four simple items and one cross item. Each simple item will implement an attribute. The cross item will implement the matrix coverage model itself.

Item instruction is of type instruction_t and is sampled from the instruction fetch monitor unit sys.instmon.

```
item instruction : instruction_t =
                   sys.instmon.instruction;
```

Item decode_path is an identifier. Each of its values is associated with a unique path through the instruction decoder. The value is captured whenever the Boolean field decode_path_traversed has the value TRUE (and instruction_interaction_e is emitted). decode_-path_traversed is assigned the value TRUE whenever event decode_-path_traversed_e is emitted. The event is emitted by the code coverage program. The item is tagged as external code coverage data whose parameters are specified by agent_options.

```
item decode_path using when =
                  decode_path_traversed,
                  external = code_coverage,
                 agent_options =
                  "verilog, path is
                   module = decoder,
                   instance = top.decode0";
```

Item exec_mode is of type bit and it samples the value on the port exec_mode_reg$ when event exec_mode_sampled is emitted.

```
item exec_mode : bit = exec_mode_reg$,
         using when = exec_mode_sampled,
```

Item `overflow` is a counter that is sampled when the Boolean field `ovfl_assertion_exec` has the value TRUE. `ovfl_assertion_exec` is assigned the value TRUE whenever event `ovfl_assertion_exec_e` is emitted by the assertion tool (normally a simulator).

```
item overflow using when =
                        ovfl_assertion_exec,
                external = assertion,
           agent_options = "ovl, name =
                         ibufovf"
```

The cross coverage item defines a full permutation of all of the values of `instruction`, `decode_path`, `exec_mode` and `overflow`. It has the name `cross__instruction__decode_path__exec_mode__overflow`.

```
cross instruction, decode_path, exec_mode,
      overflow
```

The full coverage group looks like this:

```
extend coverage_u {
  cover instruction_interaction_e is {
    item instruction : instruction_t =
                                sys.instmon.instruction;
    item decode_path using when =
                             decode_path_traversed,
                             external = code_coverage,
                            agent_options =
                             "verilog, path is
                              module = decoder,
                              instance = top.decode0";
    item exec_mode : bit = exec_mode_reg$,
              using when = exec_mode_sampled,
    item overflow using when =
                                ovfl_assertion_exec,
                           external = assertion,
                   agent_options = "ovl, name =
                     ibufovf"
    cross instruction, decode_path, exec_mode,
          overflow
  }
} // extend coverage_u //
```

Now, let's define the correlation event. The event `instruction_interaction_e` is defined as:

```
  event instruction_interaction_e is {
    @decode_path_traversed_e; [2..5];
    @ovfl_assertion_exec_e and
    @exec_mode_sampled_e
  } @clock
```

It is emitted whenever the temporal expression (TE) following the word "is" succeeds. The TE succeeds when `decode_path_traversed_e` is emitted, 3 to 6 clocks elapse and `ovfl_assertion_exec_e` and `exec_mode_sampled_e` are emitted on the same clock.

The events referenced in this TE are defined as:

```
event decode_path_traversed_e;
event exec_mode_sampled_e is cycle @clock;
event ovfl_assertion_exec_e
```

The following "on" blocks[4] record emitted events as Boolean values in their associated fields. In a sense, they translate a temporal success into the Boolean TRUE. Note the fields are reset to FALSE after an instruction interaction is captured.

```
on decode_path_traversed_e {
  decode_path_traversed = TRUE
};

on exec_mode_sampled_e {
  exec_mode_sampled = TRUE
};

on ovfl_assertion_exec_e {
  ovfl_assertion_exec = TRUE
};

on inst_interaction_capt_e {
  decode_path_traversed = FALSE;
  exec_mode_sampled     = FALSE;
  ovfl_assertion_exec   = FALSE
};
```

Event `inst_interaction_capt_e` is emitted on the clock after `instruction_interaction_e` is emitted. I also declare the three Boolean fields here:

[4] An "on" block is an ***e*** struct member.

```
    event inst_interaction_capt_e is
      { @instruction_interaction_e; [1] } @clock;

    decode_path_traversed : bool;
    exec_mode_sampled     : bool;
    ovfl_assertion_exec   : bool
```

Here, I declare the port exec_mode_reg, referenced above, and bind it to the RTL path ~/top/exmode.

```
    exec_mode_reg : in simple_port of bit
                        is instance;
      keep bind(exec_mode_reg, external);
      keep exec_mode_reg.hdl_path()
                                   == "~/top/exmode"
  } // extend coverage_u //
```

Finally, the type instruction_t is defined:

```
  type instruction_t : [
    NOP, EI,   DI,   ADD,   SUB,   AND, OR, NOT, MOV,
    JMP, CALL, RET, IRET, LOAD, STORE
  ]
```

8.2. Coverage Overlap

A verification plan usually employs multiple coverage sources for measuring verification progress. Functional coverage measures the progress through the functional requirements of the device. Code coverage measures progress in exercising the implemented RTL. Assertion coverage measures progress in activating, evaluating and executing assertions. Although each coverage metric provides a different view into verification progress, they also report some redundant information. In other words, some of the reported coverage overlaps from one metric to the other.

What if it were feasible and easy to map one coverage metric to another and maintain a coherent set of mappings between them? This would allow the verification engineer to trade off one kind of coverage for another, depending upon the performance cost of measuring these coverages. It would also allow the engineer to prune coverage goals of one metric because another metric records equivalent coverage. Let's look at code-to-functional coverage mapping in particular.

Motivated by another goal — documenting features implemented by source code — Thomas Eisenbarth and Rainer Koschke describe a procedure for mapping code coverage to features in a recent paper.[5] Features may be described by functional coverage models in the context of our application, hardware verification. This approach may be applied to a hardware design, at the behavioral or RT level, in order to bridge the gap between code coverage and functional coverage. How so?

Start with an autonomous verification environment whose generation aspect is endowed with those constraints required to conform to the device functional specification (functional constraints) and those required to implement the verification plan (verification constraints). The coverage aspect of the environment includes the functional coverage models required to define the device functional verification space. Artificially direct the environment to sequentially, and mechanically, traverse each coverage model, capturing a code coverage trace of the RTL for each model. Lastly, organize the recorded code coverage measurements in a map of the functional coverage models.

The resulting code-to-functional coverage map allows the DV engineer to optimize simulations for speed by eliminating functional or code coverage measurement and inferring that coverage from the remaining one. Alternatively, sections of code or functional coverage goals may be excised because they are captured by equivalent coverage in the other metric. For example, if RTL instrumented for code coverage substantially degrades simulation speed, create a map and infer code coverage from functional coverage. (Map coherency must be maintained as the RTL changes.) The converse scenario is also plausible.

I believe similar mapping techniques between functional and assertion coverage and code and assertion coverage are possible. The primary challenge is identifying complementary coverage measurements at the same abstraction level and maintaining mapped coverage coherency. The potential efficiency gains, coupled with complementary, high-fidelity coverage, make this a ripe area for research and development.

---

[5] "Locating Features in Source Code," Thomas Eisenbarth and Rainer Koschke, "IEEE Transactions on Software Engineering," March 2003, p. 210-224.

However, until a total coverage management solution — that includes mapping — is available for commercial use, it remains important to measure all three kinds of coverage. Each provides a unique perspective into verification progress, revealing aspects of device behavior invisible to the others.

8.3. Static Verification Coverage

Static and dynamic verification techniques are both used in the typical verification flow. Static techniques, such as model checking and hybrid approaches, are usually used at the block level because they have capacity constraints that limit them to that unit of RTL. Dynamic techniques such as simulation are used at the block, subsystem and full chip level. Using a coverage-driven verification methodology, how do we determine the coverage delivered by static methods?

The following discussion is restricted to model checking because it is much more widely employed than the other methods such as theorem proving and symbolic trajectory evaluation. Nonetheless, similar arguments may be formulated for the others. Any property specification may be described with a semantic description. For example, the PSL property:

```
property BranchDecodeRestriction3 =
    never {three_operand_ins; branch_inst};
```

has the semantic description:

> *"A branch instruction may not be decoded in the cycle immediately following the decode of a 3-operand instruction."*

This description may be associated with one or more functional coverage models through their own semantic descriptions.[6] If analysis of a property and an associated coverage model reveals that each describes a subset of the same functionality, a formal proof of the property should be recorded as full coverage in the coverage model.

Unfortunately, today this requires a completely manual process. The properties and coverage models must have rigorous semantic descriptions, to the extent possible using a natural language. These semantic descriptions must be compared and correlated through manual analysis. Nonetheless, with widespread adoption of static methods for block level verification, both

[6] Recall from chapter 4, "Functional Coverage," the first step of the top-level design of a functional coverage model is writing its semantic description.

white box and black box, unifying measurement of verification progress with dynamic methods would allow verification teams to pare simulations to those required to verify functionality not proven by static methods.

8.4. Summary

In order to precisely define device behavior, we may need to define a coverage model of heterogeneous attributes. Those attributes may be captured from functional coverage, code coverage and assertion coverage. In this chapter, I illustrated the design and implementation of a model using all three attribute sources. EDA vendors are challenged to deliver the means to define high fidelity, heterogeneous, coverage models. I also discussed the idea that coverage metrics oftentimes overlap, motivating means for mapping one type of coverage to another. Finally, the notion of coverage attributed to static verification methods, such as model checking, was discussed.

Appendix A: *e* Language BNF

```
module :=
      statement_list
;

statement_list :=
     statements
;

statements :=
      statement
    | statements ';' statement
;

statement :=
    | package_statement
    | struct_statement
    | extend_struct_statement
    | type_statement
    | extend_type_statement
    | routine_statement
    | simulator_statement
    | unit_statement
    | sequence_statement
    | method_type_statement
    | c_export_statement
;

package_statement :=
      package id
;
```

```
struct_statement :=
      package_or_null struct id like_opt
        '{' struct_member_list '}'
;

like_opt :=
    | like id
;

extend_struct_statement :=
      extend struct_type '{' struct_member_list '}'
;

type_statement :=
      package_or_null type_def id ':' scalar_type
;

extend_type_statement :=
      extend id ':' '[' enum_item_list ']'
;

routine_statement :=
      package_or_null routine id
        '(' parameter_list ')' type_opt
routine_name_opt
;

routine_name_opt :=
    | is c routine id
;

last_semi_opt :=
    | ';'
;

c_export_statement :=
      c export id c_export_opt
;
```

```
c_export_opt :=
    | '.' id '(' ')'
;

package_or_null :=
    | package
;

encap :=
    | package
    | private
    | protected
;

sequence_statement :=
      package_or_null sequence id sequence_opt
;

sequence_opt :=
    | using seq_name_pair_list
;

seq_name_pair_list :=
      seq_name_pair
    | seq_name_pair_list ',' seq_name_pair
;

seq_name_pair :=
      id '=' struct_type
;

method_type_statement :=
      package_or_null method type id
      '(' parameter_list ')' opt_return opt_event
;
```

```
opt_return :=
    | ':' type_def
;

opt_event :=
    | '@' event_ref
;

struct_member_list :=
      struct_members
;

struct_members :=
      struct_member
    | more_struct_members struct_member
;

struct_member :=
    | field_declaration
    | method_declaration
    | subtype_declaration
    | constraint_declaration
    | coverage_declaration
    | temporal_declaration
    | simulator_member
    | attribute_construct
    | cvl_declaration
;

field_declaration :=
      encap id field_type_specifier opt_instance
    | encap field_property id field_type_specifier
        opt_instance
;
```

```
field_property :=
      '!'
    | '%'
    | '!' '%'
    | '%' '!'
;

field_type_specifier :=
    | '[' expr ']' ':' list_type
    | ':' type_def
;

method_declaration :=
      encap method_name '(' parameter_list ')'
        type_opt method_specifier action_block
    | encap method_name '(' parameter_list ')'
        type_opt is empty
    | encap method_name '(' parameter_list ')'
        type_opt is undefined
    | encap method_name '(' parameter_list ')'
        type_opt is c routine id
    | encap method_name '(' parameter_list ')'
        type_opt '@' event_ref method_specifier
        action_block
    | encap method_name '(' parameter_list ')'
        type_opt '@' event_ref is empty
    | encap method_name '(' parameter_list ')'
        type_opt '@' event_ref is undefined
    | encap method_name '(' parameter_list ')'
        type_opt method_specifier foreign_opt
        dynamic c routine libname_opt
;

method_name :=
      method_id
;
```

```
parameter_list :=
    | parameters
;

parameters :=
      parameter
    | parameters ',' parameter
;

parameter :=
      id
    | id ':' type_def
    | id ':' '*' type_def
;

type_opt :=
    | ':' type_def
;

method_specifier :=
      member_specifier
    | is inline
    | is inline only
;

foreign_opt :=
    | foreign
;

member_specifier :=
      is
    | is also
    | is first
    | is only
;
```

```
libname_opt :=
    | id ':'
    | id
    | id ':' id
;

subtype_declaration :=
      encap when struct_subtype
        '{' struct_member_list '}'
;

constraint_declaration :=
      keep constraint_spec
;

list_of_constraint_spec_or_null :=
    | list_of_constraint_spec last_semi_opt
;

list_of_constraint_spec :=
      constraint_spec
    | list_of_constraint_spec ';' constraint_spec
;

constraint_spec :=
      constraint_expr
    | gen_before_subtypes '(' field_list ')'
    | reset_gen_before_subtypes '(' ')'
;

field_list :=
         id
    | field_list ',' id
;

attribute_construct :=
      attribute id id '=' attribute_expr
;
```

```
attribute_expr :=
      id
;

unit_statement :=
      package_or_null unit id like_unit_opt
        '{' struct_member_list '}'
;

like_unit_opt :=
    | like id
;

opt_instance :=
    | is instance
;

cvl_declaration :=
      cvl_method
    | cvl_call
    | cvl_callback
;

cvl_method :=
      cvl method opt_async method_name
      '(' parameter_list ')' opt_event cvl_routine
;

cvl_call :=
      cvl call opt_async method_name
      '(' parameter_list ')' opt_event cvl_routine
;

cvl_callback :=
      cvl callback opt_async method_name
      '(' parameter_list ')' opt_event cvl_routine
;
```

```
opt_async :=
    | async
;

cvl_routine :=
    | is c routine target_struct
;

target_struct :=
      id
    | id '.' id
;

hdl_path :=
      ''' hdl_pathname '''
;

simulator_statement :=
      simulator_member
    | simulator_restricted_member
;

simulator_member :=
      verilog simulator id
    | vhdl simulator id
    | verilog task hdl_path
        '(' vtask_parameter_list ')'
    | verilog function hdl_path
        '(' vfunc_parameters ')' v_size_opt
    | verilog variable hdl_path options_opt
    | verilog code expr
    | vhdl code '{' verilog_command_list
        last_semi_opt '}'
    | vhdl procedure hdl_path options_opt
    | vhdl function ''' id ''' options_opt
    | vhdl driver hdl_path options_opt
    | vhdl object hdl_path
;
```

```
simulator_restricted_member :=
      verilog time verilog_timescale
    | vhdl time vhdl_timescale
;

verilog_command_list :=
      verilog_command
    | verilog_command_list ';' verilog_command
;

verilog_command :=
      STRING_LITERAL
;

vtask_parameter_list :=
    | vtask_parameters
;

vtask_parameters :=
      vtask_parameter
    | vtask_parameters ',' vtask_parameter
;

vtask_parameter :=
      id ':' expr vtask_parameter_options_opt
;

vtask_parameter_options_opt :=
    | ':' vtask_io
;

vtask_io :=
      in
    | id
    | inout
;
```

```
vfunc_parameters :=
    | vfunc_parameter_list
;

vfunc_parameter_list :=
      vfunc_parameter
    | vfunc_parameter_list ',' vfunc_parameter
;

vfunc_parameter :=
      id v_size_opt
;

v_size_opt :=
    | ':' expr
;

verilog_action :=
      force hdl_path '=' force_rhs
    | release hdl_path
;

force_rhs :=
      expr
    | verilog_literal
;

verilog_timescale :=
      NUMERIC_LITERAL id '/' NUMERIC_LITERAL id
;

vhdl_timescale :=
      NUMERIC_LITERAL id
;

action_block :=
        '{' action_list '}'
;
```

```
action_list :=
     actions
;

actions :=
      action
    | actions ';' action
;

action :=
        e_action
;

e_action :=
    | var_action
    | assign_action
    | conditional_action
    | iterative_action
    | method_call_action
    | start_tcm_action
    | compute_action
    | return_action
    | try_action
    | check_action
    | gen_action
    | emit_action
    | time_consuming_action
    | print_action
    | verilog_action
    | debug_action
    | dut_error_action
    | do_seq_action
    | action_block
;
```

```
var_action :=
      var id type_opt init_opt
    | var id ":=" expr
;

init_opt :=
    | '=' expr
;

conditional_action :=
      break
    | continue
    | if_action
    | case_action
;

if_action :=
      if expr then_opt action_block else_part_opt
;

then_opt :=
    | then
;

else_part_opt :=
    | else action_block
    | else if_action
;

case_action :=
      case '{' case_list '}'
    | case binary_expr '{' case_list '}'
;

case_list :=
      cases last_semi_opt
;
```

```
cases :=
      case
    | cases ';' case
;

case :=
      expr colon_opt action_block
    | default
;

colon_opt :=
    | ':'
;

default :=
      default
      colon_opt action_block
;

iterative_action :=
      repeat do_opt action_block until expr
    | while expr do_opt action_block
    | for id from expr up_down binary_expr
        step_opt do_opt action_block
    | for '{' action ';' expr ';' action '}'
        do_opt action_block
    | for each iterated_type_opt itemname_opt
        indexname_opt in expr do_opt action_block
    | for each iterated_type_opt itemname_opt
        indexname_opt in reverse expr
        do_opt action_block
    | for each file itemname_opt matching expr
        do_opt action_block
    | for each line itemname_opt in file expr
        do_opt action_block
;
```

```
up_down :=
      to
    | down to
;

iterated_type_opt :=
    | struct_type
    | enumerated_type
;

itemname_opt :=
    | '(' id ')'
;

indexname_opt :=
    | using index '(' id ')'
;

do_opt :=
    | do
;

step_opt :=
    | step expr
;

try_action :=
      try
      action_block else_try_opt
;

else_try_opt :=
    | else action_block
;
```

```
check_action :=
      check name_opt that_opt expr opt_block
        dut_error_opt
    | assert expr else_error_opt
;

name_opt :=
    | '<' id '>'
;

that_opt :=
    | that
;

dut_error_opt :=
    | else dut_error_name '(' argument_list ')'
        opt_block
;

dut_error_name :=
      dut_error
    | dut_errorf
;

else_error_opt :=
    | else error '(' exprs ')'
;

method_call_action :=
      method_invocation
    | method_port_invocation
;

action_opt :=
      action_block
    | with_opt
;
```

```
expr_or_default :=
      expr
    | default
;

opt_config_param :=
    | ',' exprs
;

compute_action :=
      compute expr
;

return_action :=
      return expr_opt
;

assign_action :=
      lval_expr assign_operator expr
;
```

```
assign_operator :=
        '='
      | "+="
      | "-="
      | "*="
      | "/="
      | "%="
      | "<<="
      | ">>="
      | "&="
      | "^="
      | "|="
      | "and="
      | "or="
      | "<="
      | "&&="
      | "||="
;

gen_action :=
        gen reduced_gen_action_item itemname_opt
          keeping_opt
      | gen qualified_id itemname_opt keeping_opt
;

keeping_opt :=
        | keeping '{' constraint_list '}'
;

print_action :=
        print exprs options_opt
;

do_seq_action :=
        do when_qualified_id itemname_opt keeping_opt
;
```

```
debug_action :=
      message '(' argument_list ')' opt_block
    | messagef '(' argument_list ')' opt_block
;

dut_error_action :=
      dut_error '(' argument_list ')' opt_block
    | dut_errorf '(' argument_list ')' opt_block
;

opt_block :=
    | action_block
    ;

when_qualified_id :=
       id
    | struct_qualifier when_qualified_id
;

qualified_id :=
     path_id
   | struct_qualifier qualified_id
;

path_id :=
     id
   | id '[' expr ']'
   | me
   | path_id '.' id
   | path_id '.' id '[' expr ']'
   | path_id '.' as_a '(' type_def ')'
;
```

```
reduced_gen_action_item :=
      '.' id
    | '.' id '[' expr ']'
    | reduced_gen_action_item '.' id
    | reduced_gen_action_item '.' id '[' expr ']'
    | reduced_gen_action_item '.' as_a '(' type_def ')'
;

coverage_declaration :=
      cover id opt_cov_field coverage_group_option
;

coverage_group_option :=
      options_opt member_specifier
        '{' cover_item_list '}'
    | is empty
    | using also options opt_cover_item_list
;

cover_item_list :=
    | cover_items last_semi_opt
;

opt_cover_item_list :=
    | is also '{' cover_item_list '}'
;

cover_items :=
      cover_item
    | cover_items ';' cover_item
;

cover_item :=
      item id item_options_opt
    | item id ':' type_def '=' expr options_opt
    | transition id item_options_opt
    | cross item_name_list item_options_opt
;
```

```
item_name_list :=
      id
    | item_name_list ',' id
;

opt_cov_field :=
    | '(' expr ')'
;

item_options_opt :=
    | using options
    | using also options
;

temporal_declaration :=
      encap event id event_option
    | on id opt_defer do_opt action_block
    | encap expect_declaration
;

opt_defer :=
    | '$'
;

event_option :=
    | is temporal_expr
    | is only temporal_expr
;
```

```
expect_declaration :=
      expect id
    | expect temporal_expr dut_error_opt
    | expect id expect_specifier temporal_expr
        dut_error_opt
    | assume id
    | assume temporal_expr dut_error_opt
    | assume id expect_specifier temporal_expr
        dut_error_opt
;

expect_specifier :=
      is
    | is only
;

emit_action :=
      emit event_ref
;

start_tcm_action :=
      start method_invocation
    | start method_port_invocation
;

time_consuming_action :=
      all of action_block
    | first of action_block
    | wait
    | wait until_opt temporal_expr
    | sync
    | sync temporal_expr
    | state machine expr until_state_opt
        '{' transition_list '}'
;
```

```
until_opt :=
      | until
;

until_state_opt :=
    | until id
;

transition_list :=
      last_semi_opt
    | transitions last_semi_opt
;

transitions :=
      transition
    | transitions ';' transition
;

transition :=
      id "=>" id action_block
    | '*' "=>" id action_block
    | id action_block
;

event_ref :=
      id
    | field_access
    | primitive_expr '$'
;

temporal_expr :=
      temporal_inclusive_expression
    | temporal_expr '@' event_ref
;
```

```
temporal_inclusive_expression :=
      temporal_or_expression
    | temporal_inclusive_expression "=>"
        temporal_or_expression
;

temporal_or_expression :=
      temporal_and_expression
    | temporal_or_expression or
        temporal_and_expression
;

temporal_and_expression :=
      temporal_exec_expression
    | temporal_and_expression and
        temporal_exec_expression
;

temporal_exec_expression :=
      temporal_sampling_expression
    | temporal_sampling_expression exec action_block
;

temporal_sampling_expression :=
      temporal_eventual_expression
    | temporal_eventual_expression '@' event_ref
;
```

```
temporal_eventual_expression :=
      temporal_repeat_expr
    | eventually temporal_repeat_expr
    | not temporal_eventual_expression
;

temporal_repeat_expr :=
      temporal_unary_expr
    | '[' range_expr ']' temporal_repeat_opt
    | '[' expr ']' temporal_repeat_opt
    | '~' '[' range_expr ']' temporal_repeat_opt
;

temporal_primitive :=
      cycle
    | detach '(' temporal_expr ')'
    | true '(' expr ')'
    | rise '(' expr ')'
    | fall '(' expr ')'
    | change '(' expr ')'
    | delay '(' expr ')'
    | '(' temporal_expr ')'
    | '{' temporal_sequence last_semi_opt '}'
    | consume '(' '@' event_ref ')'
;

temporal_sequence :=
      temporal_expr
    | temporal_sequence ';' temporal_expr
;

temporal_repeat_opt :=
    | '*' temporal_exec_expression
;
```

```
temporal_unary_expr :=
      temporal_primitive
    | '@' event_ref
    | fail temporal_unary_expr
;

type_def :=
      non_port_type
    | port_type
;

non_port_type :=
      regular_type
    | list_type
;

regular_type :=
      scalar_type
    | struct_subtype
;

scalar_type :=
      id
    | enumerated_type
    | scalar_type scalar_modifier
;

enumerated_type :=
      '[' enum_item_list ']'
;

scalar_modifier :=
      '[' range_elements ']'
    | '(' scalar_unit ':' expr ')'
    | '(' scalar_unit ':' '*' ')'
;
```

```
enum_item_list :=
    | enum_items
;

enum_items :=
      enum_item
    | enum_items ',' enum_item
;

enum_item :=
      id enum_num_opt
;

enum_num_opt :=
    | '=' expr
;

scalar_unit :=
      bits
    | bytes
;

struct_type :=
      id
    | struct_subtype
;

struct_subtype :=
      struct_qualifier struct_type
;

struct_qualifier :=
      id
    | id ''' id
    | FALSE ''' id
    | TRUE ''' id
;
```

```
list_type :=
      list of type_def
    | list '(' id ':' id ')' of type_def
;

port_type :=
      io_type simple port of non_port_type
    | io_type buffer port of non_port_type
    | io_type event_ref port
    | serve_client call port of non_port_type
    | io_type method port of id
;

io_type :=
    | id
    | in
    | inout
;

serve_client :=
      id
;

constraint_expr :=
      binary_expr
;

select :=
      select '{' selection_list last_semi_opt '}'
;

selection_list :=
      selection
    | selection_list ';' selection
;
```

```
selection :=
      expr ':' expr
;

port_binding :=
        bind '(' expr ',' port_bind_target
                port_constraint ')'
        ;

port_bind_target :=
      expr
    | empty
    | undefined
;

port_constraint :=
    | ',' '{' constraint_list '}'
;

lval_expr :=
      id
    | field_access
    | primitive_expr '[' range_element ']'
    | hdl_path
    | bit_extract
    | bit_concat
    | primitive_expr '$'
;
```

```
primitive_expr :=
      lval_expr
    | me
    | literal
    | '(' binary_expr ')'
    | new_action
    | method_invocation
    | method_port_invocation
    | '[' range_elements ']'
    | cast
    | select
    | port_binding
;

new_action :=
      new
    | new struct_type itemname_opt with_opt
;

with_opt :=
    | with action_block
;

field_access :=
      primitive_expr '.' when_field_access
    | '.' when_field_access
    | when_field_access_pair
;

when_field_access :=
      id
    | when_field_access_pair
;
```

```
when_field_access_pair :=
      FALSE ''' id
    | TRUE ''' id
    | when_field_access ''' id
;

bit_extract :=
      primitive_expr '[' expr_opt ':'
        expr_opt slice_opt ']'
;

slice_opt :=
    | ':' scalar_type
;

bit_concat :=
      '%' '{' bit_elements '}'
;

bit_elements :=
      expr
    | bit_elements ',' expr
;

method_port_invocation :=
     primitive_expr '$' '(' argument_list ')'
;
```

```
method_invocation :=
      primitive_expr '.' called_method_name
        '(' argument_list ')'
    | '.' called_method_name '(' argument_list ')'
    | id_or_special_method '(' argument_list ')'
    | hdl_path '(' argument_list ')'
    | all_values '(' scalar_type ')'
    | get_all_units '(' struct_type ')'
    | primitive_expr '.' get_enclosing_unit
        '(' struct_type ')'
    | get_enclosing_unit '(' struct_type ')'
    | primitive_expr '.' try_enclosing_unit
        '(' struct_type ')'
    | try_enclosing_unit '(' struct_type ')'
    | primitive_expr '.' seq_method
        '(' type_def ')' itemname_opt
    | '.' seq_method '(' type_def ')' itemname_opt
    | seq_method '(' type_def ')' itemname_opt
;

called_method_name :=
      method_name
;

id_or_special_method :=
      method_id
;

seq_method :=
      in_sequence
    | in_unit
;

argument_list :=
    | exprs
;
```

```
cast :=
      primitive_expr '.' as_a '(' type_def ')'
    | '.' as_a '(' type_def ')'
;

range_elements :=
      range_element
    | range_elements ',' range_element
;

range_element :=
      expr
    | range_expr
;

range_expr :=
      expr_opt ".." expr_opt
;

list_elements_or_null :=
    | list_elements last_semi_opt
;

list_elements :=
      expr
    | list_elements ';' expr
;
```

```
unary_expr :=
      primitive_expr
    | now '@' event_ref
    | '{' list_elements_or_null '}'
    | '{' list_elements_or_null '}'
        '[' range_element ']'
    | '{' list_elements_or_null '}' '.' id
        '(' argument_list ')'
    | unary_operator unary_expr
    | primitive_expr unary_post_operator
    | lval_expr time_unit
    | literal time_unit
    | constraint_for_each_expr
    | text_expansion_exp
    | "<<" STRING_LITERAL
;

unary_operator :=
      not
    | '|'
    | '&'
    | '^'
    | nor
    | nand
    | nxor
    | '+'
    | '-'
    | '~'
    | '!'
;

unary_post_operator :=
      is empty
    | is not empty
;
```

```
binary_expr :=
      boolean_imp_expression
    | boolean_imp_expression '?' expr ':' expr
;

boolean_imp_expression :=
      logical_OR_expression
    | boolean_imp_expression "=>"
        logical_OR_expression
;

logical_OR_expression :=
      logical_AND_expression
    | logical_OR_expression bool_or_operator
        logical_AND_expression
;

bool_or_operator :=
      "||"
    | or
;

logical_AND_expression :=
      inclusive_OR_expression
    | logical_AND_expression bool_and_operator
        inclusive_OR_expression
;

bool_and_operator :=
      and
    | "&&"
;

inclusive_OR_expression :=
      exclusive_OR_expression
    | inclusive_OR_expression '|'
        exclusive_OR_expression
;
```

```
exclusive_OR_expression :=
      AND_expression
    | exclusive_OR_expression exclusive_operator
        AND_expression
;

exclusive_operator :=
      '^'
    | nxor
;

AND_expression :=
      in_expression
    | AND_expression '&' in_expression
;

in_expression :=
      match_expression
    | in_expression IN_operator match_expression
;

IN_operator :=
      in
    | in range_expr
    | not in
;

match_expression :=
      relational_expression
    | match_expression match_operator
        relational_expression
;

match_operator :=
      '~'
    | "!~"
;
```

```
relational_expression :=
      member_expression
    | relational_expression neq_operator
        relational_rhs
    | verilog_literal neq_operator
        member_expression
;

relational_rhs :=
        member_expression
    | verilog_literal
;

neq_operator :=
      "=="
    | "!="
    | verilog_operator
;

verilog_operator :=
      "==="
    | "!=="
;

member_expression :=
      equality_expression
    | member_expression is a struct_type
    | member_expression is a struct_type '(' id ')'
    | member_expression is not a struct_type
;

equality_expression :=
      soft_expression
    | equality_expression eq_operator
        soft_expression
;
```

```
soft_expression :=
      shift_expression
    | soft shift_expression
;

eq_operator :=
      "<="
    | ">="
    | '<'
    | '>'
;

shift_expression :=
      additive_expression
    | shift_expression shift_operator
        additive_expression
    | gen '(' gen_item_list ')' before
        '(' gen_item_list ')'
;

shift_operator :=
      "<<"
    | ">>"
;

additive_expression :=
      multiplicative_expression
    | additive_expression additive_operator
        multiplicative_expression
;

additive_operator :=
      '-'
    | '+'
;
```

```
multiplicative_expression :=
      unary_expr
    | multiplicative_expression
        multiplicative_operator unary_expr
;

multiplicative_operator :=
      '*'
    | '/'
    | '%'
    | within
;

exprs :=
      expr
    | exprs ',' expr
;

expr :=
      binary_expr
;

expr_opt :=
    | expr
;

opt_index :=
    | index '(' id ')'
;
```

```
opt_prev :=
    | prev '(' id ')'
;

constraint_for_each_expr :=
      for each itemname_opt in gen_item do_opt
                        '{' constraint_list '}'
    | for each itemname_opt using opt_index opt_prev
        in gen_item do_opt '{' constraint_list '}'
;

gen_item_list :=
      gen_item
    | gen_item_list ',' gen_item
;

gen_item :=
        primitive_expr
;

constraint_list :=
    | constraints last_semi_opt
;

constraints :=
      constraint_expr
    | constraints ';' constraint_expr
;

verilog_literal :=
      BASED_LITERAL
;
```

```
time_unit :=
        hr
      | min
      | sec
      | ms
      | us
      | ns
      | ps
      | fs
;

text_expansion_exp :=
        text begin text_list text end
;

text_list :=
        '(' expr ')'
      | STRING_LITERAL
      | text_list '(' expr ')'
      | text_list STRING_LITERAL
;

options_opt :=
      | using options
;

options :=
        option
      | options ',' option
;

option :=
        id
      | id '=' expr
      | when '=' expr
      | range_option
;
```

```
range_option :=
      ranges '=' '{' cover_ranges last_semi_opt '}'
;

cover_ranges :=
      cover_range
    | cover_ranges ';' cover_range
;

cover_range :=
      range '(' '[' range_elements ']'
        optional_range_param ')'
    | range '(' id optional_range_param ')'
;

optional_range_param :=
    | ',' exprs
;

literal :=
      STRING_LITERAL
    | NUMERIC_LITERAL
    | char_literal
    | TRUE
    | FALSE
    | NULL
    | UNDEF
    | MAX_INT
    | MIN_INT
;

id :=
      id
;
```

INDEX

Symbols
0-In Design Automation, 111.

A
AAAI, 22.
AAD, 61.
AAM, 61.
ability, xv, 3, 62, 78.
able, xi, 28, 80, 89, 120.
about, 26, 106, 110, 112, 131, 133.
above, 22, 27, 40, 46, 57, 75, 84, 92, 106, 124, 128–130, 147.
absence, 104, 114, 129.
absent, 17.
abstract, 18, 34.
abstraction, 1, 2, 18, 23–25, 31–35, 40, 67, 97, 98, 107, 126, 148.
academia, xiv.
academic, xiii.
Accellera, 99, 100.
access, 46, 49, 62.
accumulate, 90.
acidity, 117.
acknowledge, 64, 101.
across, 2, 68, 124.
action, 23, 90, 162, 164.
activate, 105, 106, 114, 130, 147.
activation, 104, 107.
active, 28, 63, 76, 89.
activity, xi, 94, 98, 123.
adapt, 114, 129, 137.
ADC, 61.
add, 21, 23, 29, 61, 97, 103, 118, 119, 124, 130, 131, 134, 141, 142, 147.
addition, 21, 23, 40, 46, 90, 94, 101, 106, 120, 130.
additional, 64, 133, 136.
address, xiii, xiv, 3, 24, 26, 27, 33, 39, 40, 45, 46, 56, 61, 62, 64, 65, 75, 87, 89, 91, 94, 109–116, 129.
ADDSUB, 23.
Adir, Allon, 76.
adjust, 137.
adopt, 112, 128, 150.
advance, ix, xiv, 109, 111.
advantage, 75, 76, 128.
AF, 58, 61, 68–70.
agent, 42, 63, 65, 66, 68, 69, 71, 73, 75.
aggregate, 134.
aggregation
    coverage hole, 135.
AH, 68.
Aharon, A., 22.
AL, 68.
algorithm, 80.
algorithmic, 1.
alias, 47.
align, 109.
alkalinity, 117.
allow, xv, 18, 73, 91, 103, 127, 133, 147, 148, 150.
alter, 49.
alternate, 80.
alternative, 148.
ambiguity, 5, 18.
ambiguous, 1, 33, 34.
Amitroaie, Cristian, xv.
analysis, ix, x, xiii, xiv, 5, 67, 77, 89, 90, 92, 95, 97, 101, 104, 106, 107, 109, 111, 114, 122, 126, 129, 131–137, 149.
analyze, 78, 90–93, 95, 97, 103, 104, 107, 109, 111, 114, 129–131, 136, 137.
antecedent, 103.
AOP, 65.
apparent, 18, 24, 75, 79, 88, 114, 126.
appear, 78, 124.
applicable, ix, 55.
application, 2, 4, 17–19, 22, 26, 33, 75, 90, 101, 113, 122, 126, 148.
apply, 2, 19, 23, 30, 32, 82, 92, 107, 111, 115, 120, 129, 131, 134, 136, 139, 148.
approach, ix, 19, 24, 25, 29, 78, 110, 112, 124, 129, 148, 149.
appropriate, 80.
approximate, 3.
arbiter, 2, 36.
arbitration, 130.
arbitrator, 36.
arc, 42, 75, 85–87, 90, 94, 95, 104, 125, 127, 128, 136.

Archer, 111.
architect, 17, 18, 61.
architectural, 21, 26.
architecture, 20, 21, 24, 26, 45, 46, 55, 59, 68, 76, 88, 110, 113, 114, 141.
area, x, 16, 76, 90, 133, 134, 149.
argue, 115.
argument, 99, 149.
arithmetic, 46, 57, 58, 61, 66, 68, 70.
arithmetic flags
    coverage model, 61.
Armbruster, Frank, xv.
ARPL, 61.
artifact, 24, 25.
artificial, 22, 148.
Asaf, Sigal, 131.
aspect, 65, 66, 137, 149.
    checking, 20, 29, 30, 63, 65, 68, 89, 98, 120, 121, 137.
    coverage, 20, 27, 29, 63, 65, 68, 98, 121, 129, 137, 148.
    design culture, 110.
    device, 19.
    functional verification, 20, 98, 111.
    generation, 20, 22, 23, 29, 30, 65, 66, 68, 112, 129, 130, 137, 148.
    stimulus, 20, 115.
    verification, 20, 21.
    verification environment, 26.
    verification plan, 21.
aspect-oriented, 27, 65.
assembly, 22, 88.
assert, 27, 28, 49, 63, 98, 100, 101, 106, 140, 166.
assertion, xiv, 5, 22, 23, 29, 30, 64, 78, 97–107, 109, 111, 113, 122, 126–128, 130, 140, 141, 144, 147, 148, 150.
    classification, 99.
    coverage, 5.
        density, 5.
        measurement, 126.
    FSM, 102.
Assertion-Based Design, xiii.
assess, 19, 31, 78, 125.
assign, 70, 88, 131, 143, 144.
assignment, 88, 93.
associate, 26, 29, 35, 37, 40, 49, 51, 62, 64, 77, 78, 88, 90, 94, 97, 104–106, 110, 126, 127, 130, 133, 135, 143, 146, 149.
assume, 35, 68, 92, 94, 127, 172.
assumption, 27, 30.
asterisk, 42, 57.
async, 159.
asynchronous, 63.
attention, 34, 76.
attribute, 5, 33, 34, 41–58, 61–64, 66, 69, 70, 75, 76, 78–80, 90, 111, 114, 121, 124, 131, 132, 134–136, 141–143, 150, 157.
attributes
    sampling input, 62.
    sampling internal, 65.
    sampling output, 64.
audience, xiv.
author, ix, 120, 135.
automate, 88, 110, 129.
automatic, ix, 75, 111, 130.
automation, 78.
autonomous, xiv, 27–30, 90, 92, 124, 148.
Averant, 78.
AX, 68, 134.
axis, 51.

B

baby steps, 26.
back-end, 1, 95, 126.
balance, 114, 128, 130.
bandwidth, 34.
base, ix, 2, 20, 22, 32, 52, 69–71, 103, 110, 112–114.
basic, xiii, xv, 27, 28, 45, 59, 85, 94.
basis, 80.
Bayesian, 120.
bearing, 31.
begin, 17–19, 22, 40, 87, 90, 124, 125, 135, 142, 191.
behave, 24.
behavior, 1, 16–19, 21, 22, 24, 25, 28, 29, 33, 34, 37–39, 45, 46, 48, 50, 55, 59, 75, 76, 78, 89, 95, 97, 101, 103, 112, 113, 120, 125, 126, 131, 132, 139, 140, 149, 150.
behavioral, 1, 16, 21, 25, 27, 40, 50, 55, 109, 121, 148.

belief, 110.
believe, 148.
below, 20, 25, 31, 37, 41–43, 49–52, 54, 59, 70, 80, 84, 98, 102, 105, 121, 123, 125, 126, 133, 142.
benefit, 89.
Bening, Lionel, 99.
Bergeron, Janick, xi, xiii, xv.
beyond, 3, 28.
BF, 17.
BH, 68.
bias, 22, 113–115, 130.
Bin, Eyal, 76.
binary, 22.
bind, 100, 147, 179.
Binyamini, Ziv, xv.
bit, 3, 40, 46–49, 51, 55, 67–69, 85, 88, 91, 94, 98–100, 114, 115, 136, 141, 143, 144, 147, 177.
BL, 68.
black, 24, 39, 67, 97, 122, 133, 134, 150.
blend, 52, 55, 117.
block, 2, 27, 63, 70, 80, 81, 105, 106, 118, 121, 125, 146, 149, 150.
body, 93.
book, x, xi, xiii–xv, 2, 4, 5, 18, 19, 23, 36, 46, 89, 102, 111, 112, 116, 120, 121, 134, 139, 140.
bool, 146.
Boolean, 2, 32, 84, 93, 100–103, 143, 144, 146.
border, 133, 134.
bound, 3, 16, 23.
boundary, 2, 23, 36, 47, 76, 83, 110, 114.
box, 24, 39, 67, 97, 122, 150.
BP, 68.
bracket, 99.
brainstorm, 45.
branch, 81, 88, 89, 92, 93, 127–129.
    coverage, 6, 82, 83, 88, 89, 92.
    instruction, 88, 149.
break, 27, 163.
bridge, 78, 148.
broad, 19, 110, 120.
broaden, 28.
broadside, 4.
browser, 103.
BSF, 61.
BSR, 61.
BSWAP, 61.
BT, 61.
BTA, 61.
BTC, 61.
BTS, 61.
bucket, 116, 119.
buffer, 76, 140, 141, 178.
bug, xiii, xiv, 1, 27, 29, 30, 76, 77, 88, 105, 110, 112, 113, 127, 128, 132, 139, 140.
build, ix, 26, 38, 55, 79, 105, 111.
built, 26, 30.
burden, 64.
bus, 2, 26, 36, 37, 40, 64, 76, 91, 94, 98, 106, 107, 134.
BX, 68, 134.
bypass, 21, 93.
byte, 46, 47, 177.

C
cache, 76.
Cadence, 99.
CAI, 103.
calculate, 104, 113.
calendar, 112.
call, 20, 22, 24, 29, 45, 46, 85, 97, 98, 134, 140, 141, 147, 158, 178.
callback, 158.
candidate, 127, 128.
capability, 4, 38, 65, 75, 79, 89, 113.
capacity, 2, 3, 149.
capitalization, 133.
capture, 4, 16–18, 21, 25, 34, 37, 39–41, 43, 44, 56, 62–64, 66, 70, 75, 76, 95, 97, 98, 124, 127, 131, 139, 141, 143, 146, 148, 150.
carry, 21, 135.
case, 17, 25, 27, 36, 44, 49, 63, 76, 77, 80, 91–93, 109, 110, 118, 120–122, 163, 164.
cast, 180, 183.
catalytic, 40.
catch, 29.
catch-22, 112.
cause, xiii, 1, 21, 22, 49, 84, 94, 105, 112,

114, 132, 139.
CDG, 113, 120.
CDV, 88, 109, 112–114, 120–123, 125.
cell, 59.
cellular, 111.
CF, 17, 58, 61, 68–70, 135.
CH, 68.
challenge, ix, xiv, 36, 116, 148, 150.
chamber, 44.
change, 24, 28, 48, 49, 58, 64, 84, 98, 106, 110, 123, 125, 129, 130, 148, 175.
channel, 18.
characteristic, 29, 116.
characterize, 2, 29, 44, 93, 114.
check, xv, 19, 24–27, 88, 100, 105, 106, 120, 125, 126, 166.
checker, 3, 4, 29, 35, 66, 77, 78, 101, 104–107, 109, 121, 122, 139.
  coverage, 6.
checking, xi, xiv, 1, 2, 19–21, 24–27, 29, 30, 63, 65, 68, 87–89, 95, 98, 101, 109, 112, 120–122, 127, 137, 149, 150.
  aspect, 121.
chip, 26, 111, 149.
choice, 24, 25, 34, 36, 67, 78, 110.
choose, 24, 33, 37, 46, 48, 55, 91, 95, 110, 123, 130.
circuit, 2.
circumstance, 23.
CL, 68.
class, 1, 34, 68, 76, 139.
classical, 28, 29.
classification, x, 31, 32, 38, 98.
classify, 31, 35, 38, 39, 97, 98, 107, 134, 135.
clause, 91.
clk, 77, 99–101.
clock, 98–101, 105, 106, 128, 145, 146.
close, 40, 42–44, 54, 113, 139.
closure, xiv, 116, 127–129.
coalesce, 134.
coarse, 94, 130.
code, ix, 30, 63, 68–71, 79, 83, 88–90, 92, 98, 106, 110, 111, 122, 125–127, 130, 142, 143, 147, 148, 150, 159.
  coverage, xiv, 6, 21, 22, 30, 35, 36, 38, 40, 75, 78–82, 85, 87–91, 95, 103, 106, 109, 111, 113, 120, 122, 124, 125, 140, 141.
  e, 26.
  error-correcting, 17.
  packaging, 26.
  RTL, 106, 124.
  verification, 26.
  writing, 27.
code coverage density, 6.
code segment, 47.
coherent, 147, 148.
coincidence, 17.
coincident, 17.
colleague, xv.
collision, 130.
colloquial, 28.
color, 133.
column, 42, 56, 59.
combination, 50, 52, 75, 107, 111.
combinational, 2.
combine, 3, 49.
combustion, 40.
comment, 80.
commercial, 26, 76, 77, 89, 111, 116, 139, 149.
common, ix, xiii, 4, 5, 26, 35, 45, 72, 76, 88, 99, 101, 109–112, 131, 135, 139.
commonality, 134.
communicate, 5.
communication, 5, 18, 105, 114.
compact, 75.
comparable, 24, 55.
comparative, 18.
compare, 2, 18, 20, 24–26, 77, 94, 97, 112, 120, 125, 149.
comparison, 137.
compatibility, 110.
compilation, 90.
compiler, 18.
complement, 122.
complementary, 148.
complete, xi, xiii, 2, 3, 5, 39, 50, 61, 64, 67, 75, 83, 84, 92, 107, 112, 120, 130, 137, 149.
completeness, 2, 3, 47.

completion, 112, 123.
complex, ix, xiv, 2, 24, 27, 55, 67, 88, 95, 111.
complexity, ix, 3, 30, 55, 90, 91.
compliant, 65.
component, 26, 30, 33, 63, 76, 79, 120, 121.
compose, 16, 20, 23, 27, 29, 30, 35–37, 48, 52, 53, 56, 57, 62, 65, 79, 86, 93, 107, 113, 114, 134–137, 140.
composite, 101.
composition, xv, 22.
comprehensive, ix, x, 2.
comprise, 76.
compromise, 62, 78.
compute, 167.
computer, xv, 51.
computing, 25.
conceive, xv, 16.
concept, xiv, xv, 18, 38, 137.
concurrent, 89, 98, 101, 105–107, 127.
condition, 23, 36, 41, 43, 47, 67, 76, 77, 83–85, 91–95, 105, 110, 113, 114, 116.
  coverage, 6.
conditional, 92, 93, 106, 127, 128.
conduit, 133.
cone, 94.
configuration, 45, 80, 91, 94.
configure, 91.
confirm, 106, 122.
conform, 19, 24, 101, 148.
conjunction, 48.
consecutive, 101.
consequence, 26.
conserve, 80.
consistency, 118.
constant, 94.
constituent, 67, 93.
constrain, 28, 115, 117.
constraint, ix, 2, 3, 23, 27, 113–118, 120, 129, 130, 135, 137, 140, 148, 149, 190.
construct, 29, 35, 50, 75, 83, 91, 93, 100, 111, 120, 130, 139.
consume, xiii, 175.
consumption, 76.
contain, 29, 46, 47, 63, 92, 103, 117.
contemporary, ix.
context, 17, 18, 23, 25, 36, 88, 109, 122, 148.
continual, 24, 64.
continue, ix, 163.
contrast, 29, 104, 109, 126.
contribute, 29, 56, 57, 112.
contribution, 65, 78.
control, xi, 40, 45, 46, 48, 49, 52, 67, 76, 80–84, 87, 89, 92–95, 98, 106, 107, 128, 141.
controller, 2.
controversy, 28.
conventional, 16, 113, 114, 137.
converse, 148.
convert, 132.
converter, 40.
Convex Computer Systems, xv.
coordinate, 23, 64.
copy, 29, 44, 103.
core, 101.
corner, 36, 109, 110, 114, 122.
cornerstone, 122.
correct, xi, 17, 18, 139.
correction, xv.
corrective, 4.
correctness, 112.
correlate, 51, 56, 58, 61, 89, 95, 141, 149.
correlation, 41–43, 49, 51, 56, 58, 61, 66, 67, 69, 71, 73, 75, 120, 142, 143, 145.
correspond, 91.
corresponding, 47, 51, 55, 61, 70, 89, 93, 94, 116, 121, 123, 125, 139.
cost, xiii, 24, 148.
count, 80, 82, 83, 90, 94.
counter, 3, 4, 81, 90, 100, 104, 144.
counterpart, 103.
couple, 92, 106, 148.
course, xiv, 63.
cover, 43, 66, 69, 71–74, 80, 101, 118, 136, 142, 144, 170.
coverage, ix–xi, xiii–xv, 2, 4–6, 19–24, 26–45, 50–58, 61–73, 75–95, 97, 98, 101–107, 109–116, 118–132, 134–137, 139–144, 147–150.
  analysis, 6, 132, 133.
  aspect, 121.

assertion, 5, 21, 22, 30, 90, 97, 109, 111, 120, 122, 130.
branch, 6, 82, 83, 88, 89, 92.
checker, 6.
closure, 6.
code, xiv, 6, 21, 22, 30, 35, 36, 38, 40, 75, 78–82, 85, 87–91, 95, 103, 106, 109, 111, 113, 120, 122, 124, 125, 130, 140, 141.
condition, 6.
cross, 8.
database, 6.
density, 6.
expression, 9.
fidelity, 139.
FSM, 86.
arc, 87.
sequential arc, 87.
functional, 9, 21, 22, 30, 39, 90, 109, 120, 122, 124, 130, 140.
goal, 7.
assertion, 112.
code, 112.
functional, 112.
group, 7.
hierarchical model, 52.
hole, 120.
hole aggregation, 135.
hybrid model, 53.
implicit, 10, 80.
input, 10.
internal, 10.
item, 7.
language of, 5.
line, 11, 80.
matrix model, 51.
measurement, 7.
assertion, 126.
measuring, 31.
merge, 11.
metric, 7.
branch, 95.
condition, 95.
event, 95.
line, 95.
statement, 95.
model, 7, 41, 42, 51–53, 56–58.
abbreviated, 57.
arithmetic flags/instruction, 61.
GPR/arithmetic flags, 58.
structure, 71.
virtual-8086 mode, 56, 57.
output, 11.
path, 11.
point, 7.
report, 7.
sequential, 11.
space, 8.
spaces, 38.
statement, 11.
structural, 35, 38.
taxonomy, 32.
temporal, 11.
toggle, 12, 85.
user interface, 132, 133.
CPU, 33, 68.
CR, 40.
CRC, 33.
create, x, 23, 49, 62, 90, 92, 134, 148.
creep, 17.
cripple, xiii.
criteria, 29, 46.
critical, 26.
cross, 43, 55, 69, 119, 125, 136, 143, 144, 170.
coverage, 8.
crucial, xi, 128.
CS, 134.
culture, 110.
current, 23, 68, 75, 104, 110, 114, 129, 130.
cursor, 133.
curve, 122–124.
cvl, 158.
CX, 68.
cycle, 3, 4, 17, 22, 23, 27–30, 48, 76, 89, 98, 102, 106, 111, 114, 124, 127–129, 131, 137, 139, 141, 145, 149, 175.

D
DAC, 78, 120, 131.
damper, 40–44.
data, xiv, 23–26, 28, 35, 37, 39, 46, 49,

61–66, 68, 76, 77, 80, 89–92, 94, 103, 105, 114, 120, 121, 126, 130, 131, 133, 134, 137, 141, 143.
data coverage, 8.
database, 90, 103, 104.
date, 18, 29, 36, 89, 112.
deactivate, 91.
deassert, 98, 100.
decide, 48.
decision, 51, 95, 112.
declarative, 98, 99.
declare, 23, 43, 66, 68, 69, 71, 73, 146, 147.
decode, 45, 141, 149.
decoder, 45, 140, 141, 143, 144.
decompose, 100.
decomposition, 80, 104.
decouple, xv, 113, 120, 137.
deduce, 4.
deep, 3, 36, 55, 133.
default, 90, 91, 164, 167.
definition, 1, 5, 15, 18, 19, 29, 31, 79, 110, 136, 141.
degradation, 76, 80.
degrade, 90, 148.
degree, 40.
delay, 23, 175.
delegate, 22.
delimit, 99.
deliver, 18, 28, 76–78, 112, 113, 140, 149, 150.
delivery, 17.
demonstrate, 1, 15, 18, 24, 27, 104, 110.
denominator, 127, 128.
density, 102, 128, 129, 137.
depend, 1, 3, 23, 39, 46, 79, 90, 126, 148.
dependency, 21, 40, 50, 62.
dependent, 46.
depict, 62–64.
deploy, xv, 22, 124.
depth, 1, 55, 95, 107, 122.
derive, xiv, 18, 20, 23, 32–34, 36, 37, 39, 100, 104, 110, 113, 114, 121, 135.
description, 2, 18, 32, 33, 35, 41, 45, 54, 57, 77, 80, 107, 135, 141, 149.
design, ix, xiv, 1, 18, 24, 26, 29, 30, 37, 39–41, 44, 45, 50, 54, 55, 57, 61, 64, 76–79, 88–91, 95, 97, 99, 102, 103, 106, 107, 109–114, 120, 121, 124–127, 130, 131, 136, 137, 139, 140, 148, 150.
    ASIC, ix.
    complex, ix.
    detailed, 21, 22, 44, 61, 65–67, 75, 76, 78, 107, 109, 121, 141, 142.
    hardware, xiv, 18.
    logic, xiii–xv, 17.
    object-oriented, 75.
    reusable, 26.
    SoC, ix.
    top-level, 44, 61, 62, 66, 67, 78, 107, 114, 135, 141, 149.
    verify a, 4, 19.
design architecture, 37.
design choice, 34, 36.
design cycle, 17.
design errors, 5.
design intent, xiv, 15–18, 97, 126, 127.
design process, 44.
design space, 19.
design specification, 18, 20, 21, 24, 26, 31, 33–36, 39, 95, 97, 98, 111, 113, 114, 123, 130.
design verification, xiii–xv, 1, 5, 22.
designate, 101.
designer, xiv, 1, 16, 18, 27, 34, 36, 45, 75, 91, 98, 101, 125, 132, 141.
desire, 40, 110, 115.
destination, 33, 57, 58, 68, 69.
detach, 175.
detail, ix, 5, 21, 22, 24, 25, 32, 34–37, 39, 40, 44, 61, 65, 66, 75, 76, 78, 90, 95, 97, 107, 109, 121, 126, 133, 134.
detect, 2, 21, 29, 30, 35, 67, 100, 101, 104, 105.
detection, 23.
detector, 130.
determine, xiii, 2, 19, 37, 40, 49, 54, 61, 62, 77, 78, 83, 89, 92–94, 104, 114, 120, 128, 134, 135, 149.
develop, xv, 22, 28, 29, 76, 77, 88, 100, 123.
developer, 25, 26, 45, 84.
development, ix, xv, 27, 67, 88, 107, 123,

126, 149.
deviate, 21, 28.
deviation, 21.
device, xiii, xiv, 1–4, 8, 16–40, 44–46, 50, 51, 55, 61–64, 66–68, 75, 76, 78–80, 89, 90, 95, 97, 98, 101, 103–106, 109, 111–115, 120–123, 125, 127–132, 139–141, 147–150.
DH, 68.
DI, 68, 147.
diagnose, 27, 139.
diagram, 15, 16, 121.
dictate, 26, 39, 67, 104, 112.
dictionary, 29.
differ, 18, 28, 29, 37, 134.
difference, 19, 30.
different, 31, 38, 56, 70, 73, 91, 97, 107, 127, 129, 139, 140, 147.
difficult, ix, 19, 67, 91.
digest, 126.
digital, xiii–xv, 17.
dimension, 31, 45, 51, 98, 135.
dimensional, 34, 51.
direct, 4, 27, 29, 30, 40, 49, 52, 64, 76, 92, 109, 110, 113–116, 118, 120, 128–130, 137, 148.
direction, xi, xv.
directive, 100.
director, xi, 18.
disable, 101.
discipline, ix.
discover, 1, 4, 19, 78, 88, 106, 120, 124, 125, 131, 136, 139.
discovery, 128.
discrepancy, 1, 125.
disk, 80.
dispatch, 29, 30.
dispense, xiv.
display, 91, 103, 133.
dissect, 129.
distance, 134.
distant, 3.
distinct, 63, 80.
distinction, 95, 113.
distinguish, xiii, 15, 18, 34, 78, 90, 97, 101, 107, 130, 136, 140.
distribute, 24–26, 65, 103.
distribution, 102, 114, 115, 119, 129, 130.
divide, 23, 44.
DL, 68.
document, 18, 38, 48, 59, 148.
documentation, 90.
domain, 28, 33, 37, 76, 77, 130.
Donnan, Russ, xv.
Dorfman, B., 22.
draft, 26, 111.
draw, 21, 36, 46, 51, 106, 116, 140.
drawback, 75.
drill, 133.
drive, 92, 114, 123.
driven, ix, x, xiv, 22, 23, 28–30, 63, 77, 88–90, 109–114, 120, 122, 126, 127, 137, 149.
driver, 159.
duality, 115.
duplicate, 80.
DUT, 8, 19, 120.
DUV, 1, 9, 19, 24, 26, 30, 38, 62, 64, 76, 103, 112, 141.
DV, 27, 37, 148.
DX, 68.
dynamic, 1–4, 18, 20, 22, 65, 77, 104, 112, 113, 129, 137, 149, 150, 155.

E
e, 9.
EAX, 47, 68.
EBP, 68.
EBX, 47, 68.
ECX, 68.
EDA, 99, 110, 111, 150.
edge, 52, 69, 98–101, 105, 114, 120.
EDI, 68.
EDX, 68.
effect, 33, 88, 89, 95.
effective, x, 19, 45, 129, 139.
efficiency, 127–129, 137, 148.
effort, xi, xiii, 24, 53–55, 67, 111, 126.
EFLAGS, 46, 48, 50, 51, 55, 56, 58, 59, 67–70, 135.
  instruction effects, 59.
EH, 17.

EI, 147.
EIP, 46, 47, 50.
elapse, 145.
elastic, 77.
electrical, xiv.
element, 32, 35, 36, 46, 115.
embed, 97.
emit, 43, 44, 67, 69–71, 77, 84, 119, 127, 128, 142–146, 172.
empty, 155, 170, 179, 184.
enable, 49, 80, 91, 103, 126, 135.
encode, 46, 80, 98.
encoder, 80.
encoding, 46, 80.
endian, 46.
endproperty, 101.
endunit, 100.
engine, 1, 104.
engineer, ix, x, xiv, 18, 19, 27, 33, 36, 37, 45, 51, 55, 57, 76, 77, 79, 85, 97, 110, 112, 120, 121, 128, 140, 147, 148.
engineering, xiv, 18, 110, 111, 114, 148.
English, 28, 35, 45, 107.
enter, 55, 85, 107.
entity, 105.
entry, 59.
enumerate, 46, 55, 68, 71, 109, 110, 118.
envelope, 113.
environment, xiv, xv, 2, 19–24, 26–32, 44, 61, 62, 65–68, 75, 77, 88–90, 92, 98, 103, 110–115, 120–122, 124, 125, 129, 130, 136, 137, 139, 141, 148.
Envisioning Information, 134.
equal, 129, 135.
equation, 76, 85, 86, 95, 128.
equivalence, 1, 2.
equivalent, 2, 78, 128, 148.
erroneous, 88, 135, 136.
error, 1, 5, 17, 23, 24, 30, 88, 91, 103, 106, 107, 114, 115, 132, 139, 166.
escape, xiii, 88.
ESI, 68.
ESP, 68.
establish, ix, 122.
ethernet, 33.
evaluate, 84, 92, 93, 98, 101–103, 106, 107, 127, 128, 141, 147.
evaluation, 93, 102, 104, 149.
event, 9, 22, 43, 44, 48, 61, 63, 67, 69–71, 73, 75, 77, 79, 84, 94, 97–100, 102, 104–107, 118, 119, 127, 128, 142–146, 171.
eventual, 98, 112.
eventually, 175.
evidence, 106.
evolve, 22.
exact, ix, 97.
examination, 106.
examine, 17, 19, 21, 23, 28, 37, 55, 61, 78, 79, 90, 92–94, 102–104, 106, 112, 121.
except, 61, 92, 106.
exception, 29, 55, 139, 140.
excessive, 76.
exchange, 56.
excise, 148.
exclude, 46, 48, 88, 90, 91.
exclusion, 16.
exclusive, 84, 93.
exec, 77, 144, 174.
executable, 2, 25, 111.
execute, 2, 18, 46, 47, 49–51, 68, 80, 81, 83, 88, 91–94, 98, 101–103, 105–107, 117, 126–130, 141, 147.
execution, 33, 46, 49, 55–57, 81, 82, 88, 89, 98, 129, 130, 134, 140, 141.
exercise, ix, 3, 19, 22, 23, 27, 29, 35, 38, 80, 83, 89, 91, 94, 95, 112, 113, 122, 125, 127, 128, 130, 147.
exhaustive, 3.
exhibit, 22, 24, 28, 98, 112.
exist, 76, 115.
existing, 65, 77, 110, 118, 131, 142.
expect, xv, 2, 18–20, 25, 91, 98, 104, 118, 120, 132, 172.
explanation, 18, 103.
explicit, 31, 33, 35–39, 57, 84, 112, 123, 140.
explicit coverage, 9.
exploration, 2, 113.
exploratory, 130, 131.
explore, 1, 3, 19, 32, 39, 78, 89, 113, 120.
export, 152.

expose, xiv, 1, 21, 29, 30, 77, 114, 115, 139, 140.
expression, 9, 32, 67, 75, 77, 80, 84, 92, 93, 97, 99, 100, 102, 103, 127, 128, 145.
extend, 27, 42–44, 65, 66, 68–74, 103, 116, 118, 119, 131, 142, 144, 147, 152.
extension, 63, 139.
external, 64, 79, 103, 140, 143, 144, 147.
extract, 31, 33, 35, 36, 45, 75, 79, 85, 95, 103.
extraction, 75, 79, 95, 103.
extraneous, 55.

F
facilitate, 111, 132.
facility, 63, 76.
fact, xi, 92, 106.
factor, 112, 123, 128.
fail, 3, 27, 99, 102–104, 126, 141, 176.
failure, 27, 103, 112.
false, 83, 84, 92, 93, 146, 177, 181, 192.
familiar, xi, xiv, xv, 45, 76.
fault, 140.
feature, xi, xiii, 17, 29, 38, 45, 46, 67, 89, 113, 148.
feedback, xv, 114, 129, 130, 137.
fetch, 27, 50, 98, 140, 141, 143.
fidelity, xiv, 22, 40, 53–55, 78, 87, 120–122, 124, 130, 131, 136, 137, 139, 140, 148, 150.
field, 23, 28, 43, 45, 46, 61, 68, 71, 73, 75, 77, 114, 116–118, 143, 144, 146.
FIFO, 3, 105.
file, 17, 21, 22, 27, 65, 91, 103, 117, 164.
fill, xi, 9, 91, 115, 119, 130.
filter, 35, 37, 91, 95, 126.
Fine, Shai, 120.
finite, xiv, 34, 36, 39, 75, 79, 85, 104, 125, 127.
flag, 21, 23, 46, 48, 50, 57–59, 61, 66, 68, 70, 135.
flaw, 19, 104, 125.
flexible, 23.
flip-flop, 3.
float, 88.
flow, ix, 22, 29, 36, 76, 77, 81–83, 92, 93, 106, 107, 110, 112, 114, 126, 141, 149.
fluid, 114.
flux, 123.
focus, 90.
footprint, 140.
force, 161.
foreign, 156.
form, ix, x, 84, 100.
formal, 1, 3, 77, 78, 97, 102, 104, 149.
formula, 2.
formulate, 149.
ForSpec, 100.
forward, 3.
Foster, Harry, x, xiii, xv, 99.
foundation, ix, 100, 109.
foundry, 17.
fraction, 102, 104, 112.
frame, 33.
frequency, 49, 102, 106.
frequent, 2, 24, 48–50, 64, 76, 105, 106, 130.
frozen, 28, 64.
FSM, xiv, 75, 76, 78, 79, 82, 85, 86, 90, 91, 94, 95, 103, 105, 125, 127–129.
    assertion, 102.
    coverage, 86.
        arc, 87.
        sequential arc, 87.
full
    assertion coverage, 109.
    behavior, 39.
    chip level, 149.
    condition coverage, 95.
    correlation time, 67.
    coverage, 22, 28, 127, 134, 149.
    coverage group, 144.
    design space, 19.
    functional coverage, 22.
    input coverage, 112.
    line coverage, 80.
    model, 3.
    permutation, 54, 58, 141, 144.
    regression, 30.
    scope, 39, 109.
    set, 3, 28, 115.
    size, 55.

space, 114.
statement, 92.
test suite, 88.
function, 2, 29, 111, 122, 125, 159.
functional, xi, xiii, xiv, 1–4, 15–24, 26, 29–31, 33, 37–40, 44, 63, 75–79, 88, 89, 95, 97, 98, 101–103, 106, 107, 109, 111–116, 120–128, 130–132, 135, 137, 139, 140, 147–150.
functional coverage density, 10.
functional coverage development, 124.
functional specification, 35.
functionality, ix, 18, 27, 95, 104, 125, 128, 149, 150.
fundamental, 29, 109.
future, 89, 110.

G
gap, 40, 148.
gate, 1, 2, 18.
gating, 22, 105.
gen, 115, 116, 168, 188.
general, ix, xiii, 23, 24, 46, 47, 50, 57, 58, 63, 69, 94.
generalize, 120.
generate, ix, 20, 22–24, 27, 114, 116–120.
generation, xiv, 20–23, 27, 29, 30, 65, 66, 68, 88, 109, 110, 112–116, 118, 120, 127, 129, 130, 137, 148.
generator, ix, 22, 62, 76, 105, 112, 130, 132, 141.
generic, 63.
genesis, 18.
goal, 22, 28–30, 77, 90, 106, 109–111, 113, 118, 129–131, 135, 137, 148.
Gofman, E., 22.
GPR, 46, 47, 58, 68–70.
GPR/arithmetic flags
coverage model, 58.
grade, 10, 88.
grammar, xv, 35.
grammatical, 36.
grant, 23.
graph, 52.
gratuitous, 17.
grid, 130.
groff, xv.
group, 40, 41, 43, 55, 63, 65, 66, 69–73, 75, 88, 118, 119, 135, 136, 139, 142–144.
guidelines, 30, 66.

H
Hamming, 134.
hardware, xiv, 2, 18, 19, 26, 32, 35, 80, 109, 128, 148.
HDL, 79, 113.
heterogeneous, xiv, 26, 136, 140, 150.
hierarchical, 45, 51–55, 58, 59, 70, 75, 76, 78, 131, 136, 137.
coverage model, 52.
hierarchy, 75, 89, 136.
hit, 10, 80, 90, 140.
HLVL, 10, 23, 27, 28, 98, 109, 113, 114.
hole, 10, 88, 91, 101, 105, 106, 111, 114–116, 119, 120, 125, 129–132, 134–137.
Hollander, Yoav, xv, 5.
hosting struct, 71.
hybrid, xiv, 1, 3, 4, 22, 51, 52, 55, 70, 75, 78, 131, 136, 137, 140, 142, 149.
coverage model, 53.

I
IA-32
execution mode, 50.
IBM, 22, 76, 100, 120, 131.
identical, 29.
identification, 48.
identifier, 90, 117, 143.
identify, ix, xi, 44, 45, 50, 64, 75, 85, 86, 90, 112, 117, 135, 136, 143, 148.
idle, 102.
IEEE, 2, 5, 76.
ignore, 2, 43, 55, 72–74, 136.
illegal, 55, 91.
immediate, 23, 75, 101, 102, 149.
implement, 2, 16, 17, 19, 26, 27, 30, 38, 42–45, 54, 55, 61, 63–65, 67, 68, 70, 71, 73, 75, 77, 79, 90, 95, 98–102, 107, 109–111, 113, 114, 124–126, 136, 142, 143, 147, 148.
implementation, xiv, 5, 15, 17–21, 23–25,

30–39, 51, 53, 55, 61, 63, 65, 67, 68, 75, 76, 78–80, 89, 95, 102, 106, 107, 109, 110, 112, 113, 121, 123, 124, 126, 127, 130, 139, 140, 142, 150.
implication, 78, 102, 137.
implicit, 31, 32, 35, 36, 38, 79, 95, 124, 140.
imply, 2, 31, 40, 55, 119, 136.
impose, 3, 28, 39, 64.
impossible, 55, 132.
improve, xiv, 69, 87, 124, 131, 139.
improvement, 128.
incomplete, 18.
increment, 40, 90, 93.
incremental, 18, 28–30.
incubation, 27.
independent, 32, 80, 93, 105, 106, 129.
index, 165, 189.
indistinguishable, 25, 139.
individual, 55, 134.
infer, 31, 36, 104, 116, 120, 148.
infinite, 139.
influence, 46, 59, 112, 120.
information, 2, 17, 18, 37, 65, 75, 91, 107, 121, 126, 133, 147.
inheritance, 75.
initial, 4, 29, 114, 130.
initiate, 98.
inject, 18, 27, 29, 62.
inline, 156.
innovation, xv.
innovative, 22.
input, 1, 3, 4, 25–29, 37–39, 45, 55, 62–66, 68, 69, 101, 104–106, 112–116, 118, 120, 121, 130–132, 139.
insensitive, 88.
insert, 29, 81, 90, 91, 104, 106, 139.
insight, 35, 38, 55, 126, 127, 137.
instance, 19, 34, 66, 68, 70, 73–75, 79, 80, 89, 91, 95, 103, 110, 125, 143, 144, 147, 158.
instantiate, 79, 99.
instruction, 21–23, 26, 27, 34, 45–47, 49, 50, 54, 58, 59, 61, 66–68, 88, 98, 110, 134, 140, 141, 143, 144, 146, 149.
    effects, 59.
instrument, 77, 79, 89–91, 103, 148.
instrumentation, 89, 90, 103.
integrate, x, xi, xiv, 79, 111.
Integrated System Design, 19.
integration, 62, 66.
integrity, 17.
Intel, 45, 100.
intend, xi, 1, 16, 17, 33, 45, 67, 94, 97, 130–132.
intent, xiv, 15–18, 97, 101, 126, 127.
intentional, 18.
interact, 49, 55, 57, 61, 105, 124.
interaction, xi, 50, 146.
interactive, 4.
interest, 4, 57, 79, 85, 89, 133, 134.
interesting, xi, 28, 79, 93.
interface, 26, 31, 36–39, 62–67, 76, 78, 87–89, 91, 95, 103, 105, 106, 120, 132–134.
    properties, 105.
intermediate, 131.
internal, 24, 37, 38, 40, 55, 64, 65, 67, 68, 76, 87, 112, 114, 116, 120–122, 130–132.
interpret, xiv, 90, 129.
interpretation, xiv, 129.
interrelate, 75.
interrelationship, 39, 54.
interrupt, 67, 82, 140.
interval, 76, 77.
Introduction, 1.
intuition, 114.
invalid, 23, 55, 131, 132.
invalidate, 104.
invent, 18, 33, 36, 37, 99, 110.
inverse, 49.
invert, 51, 93.
invest, 88, 95.
investigate, xiv, 67.
investigation, 75.
invisible, 149.
IP, 26, 30.
IRET, 141, 147.
irregular, 52, 55.
ISA, 26, 68–70, 88, 110.
item, 43, 69, 71–74, 116, 118, 119, 136, 143, 144, 170.

iteration, 17, 83, 93.
iterative, 30.

J
James, Peet, 109.
Jasper Design Automation, x.
Java, 68.
JMP, 134, 141, 147.
Johnson, Vern, xv.
judgement, 114.

K
Kantrowitz, Michael, xv.
keep, 27, 47, 90, 115, 116, 118, 147, 157.
keeping, 115, 116, 168.
Kenville, Tom, xv.
kind, xiv, 2, 22, 31, 32, 35, 38, 46, 68, 75, 76, 78, 97, 101, 105, 107, 120, 122, 131, 140, 147, 149.
knowledge, ix, 38, 120.
Krolnik, Adam, xiii, xv.

L
label, 16, 42, 51, 63, 92, 106, 124, 133.
Lacey, David, xiii.
language, ix–xi, xiii, 1, 2, 4, 5, 18, 19, 22, 23, 30–36, 45, 61, 65, 67, 76, 80, 88, 95, 97, 99, 100, 103, 107, 109, 111, 128, 139, 142, 149.
Lapides, Larry, xv.
latch, 2, 3, 76.
latency, 34.
latent, 110.
layer, 100, 101, 106, 107.
layout, 17.
leaf, 34, 52, 53, 128.
Leibowitz, M., 22.
lesson, 89, 106.
level, ix, 1, 2, 5, 18, 21–25, 31, 32, 34–39, 42, 44, 45, 52, 61, 62, 66, 76, 78, 80, 91, 97, 98, 100, 101, 105, 107, 109, 113, 126, 127, 136, 148–150.
leverage, 121.
library, x, 99, 103.
Lichtenstein, Y., 22.
limit, xv, 2, 104, 149.
limitation, 3.
line, ix, 22, 42, 43, 49, 55, 80, 81, 91, 92, 94, 98, 100, 102, 114, 124, 127–129, 135, 164.
linear, 123.
list, ix, xi, 41, 56, 100, 109, 110, 114, 115, 117, 178.
liveness, 97, 98.
locality, 76, 134.
localize, 125.
logic, xiii–xv, 1, 2, 17, 18, 23, 49–51, 57, 76, 79, 80, 91, 94, 95, 97, 100, 104, 105, 107, 129, 130.
logical, 2.
loop, 83, 93, 113, 117.

M
machine, xiv, 18, 34, 36, 39, 75, 79, 85, 125, 129, 172.
machinery, 20, 24.
machines, 75, 85, 104, 127.
macro, xv.
magnitude, xiv, 106.
mainframe, xiv.
maintenance, 24, 64.
malfunction, 140.
Malka, Yossi, 22.
manage, 111.
management, 40, 46, 56, 70, 88, 103, 110, 149.
manager, 110.
manual, 5, 26, 45, 47, 59, 111, 114, 129, 130, 137, 149.
map, xi, 35, 44, 61, 64, 76, 77, 111, 130, 147, 148.
mapping, 61, 132, 147–150.
Marcus, Eitan, 131.
Martin, Marshall, xv.
matrix, 45, 51–55, 58, 70, 75, 78, 130, 131, 134–137, 141, 143.
maximal, 129.
maximize, xiv, 26, 127–129, 137.
maximum, 129.
MBTG, 22.
MCDC, 84.
measurable, 110.

measure, ix, xi, xiii, xiv, 19, 22, 31, 35, 38, 39, 66, 76, 78–80, 88–90, 94, 95, 97, 101–103, 107, 109, 111, 112, 125, 127–130, 135, 137, 139, 147–149.
measurement, ix–xi, xiii–xv, 2, 5, 21, 22, 27, 39, 64, 65, 68, 77, 78, 88, 90, 91, 103, 109, 111–113, 122, 124, 126, 129–131, 137, 148, 150.
coverage
assertion, 126.
mechanism, 20, 23, 35, 36, 103.
member, 68, 118, 146.
memory, 80.
message, 18, 91, 169.
messagef, 169.
method, x, xiii, 1–4, 18, 23, 26, 77, 78, 118, 119, 140, 149, 150, 153, 158, 178.
methodology, ix, x, xiv, 26, 30, 88, 109, 110, 112, 113, 120, 122, 123, 149.
metric, ix–xi, xiii, xiv, 2, 22, 31–39, 79–85, 87–92, 95, 109, 123–129, 134, 136, 140, 147, 148, 150.
branch, 80, 89.
condition, 80, 89.
coverage, 32.
event, 80, 89.
FSM, 80, 89.
line, 80, 89.
statement, 80.
toggle, 80, 89.
microarchitecture, 18, 21, 34, 37, 39, 97.
microprocessor, 88.
milestone, 109.
minimize, xiv, 64, 111, 129.
minimum, 80, 94, 119.
misbehavior, 29, 35, 76, 88.
misinterpretation, 1, 5.
misunderstand, 132.
mitigate, 90.
mode, 33, 45, 46, 49, 51, 55–57, 67, 80, 134, 139–141, 143, 144.
model, ix, x, xiv, xv, 1–5, 11, 22, 24–26, 34, 37, 39–45, 50–59, 61, 62, 64–68, 70, 71, 73, 75–81, 87–89, 95, 98, 106, 107, 109, 111–115, 121, 123–126, 129–132, 134–137, 139–143, 148–150.
hybrid, 139.
modeling, 33, 40, 100, 101.
modification, 4.
modify, 131.
ModR/M, 46.
modular, 63.
modularity, 69.
modulate, 40.
module, 79, 80, 89, 91, 95, 99, 100, 110, 121, 130, 143, 144, 151.
moment, 51, 67, 76, 94.
monitor, 23, 25, 42, 49, 62–68, 79, 89, 99, 100, 105, 116, 121, 137, 141, 143.
motto, 27.
MOV, 67, 141, 147.

N
narrow, 27.
natural, 18, 32, 33, 35, 36, 149.
nature, 1, 18, 46, 48, 55, 77.
necessary, xi, 37, 75, 80, 90, 92, 95, 104, 109, 113–116, 120, 126, 130, 134, 137.
negation, 43.
negative, 100.
negedge, 100, 101.
net, 65.
network, 120.
new, 180.
node, 52, 53, 121.
NOP, 141, 147.
nor, 76, 184.
notation, 46.
Noy, Amos, 23.
nuance, xiii, 78.
null, 192.
nxor, 184, 186.

O
object, 65, 68, 75, 110, 159.
objection, 109–112.
objective, 17, 125.
observability, 122.
observation, 67, 88, 89, 126, 129, 132.
observe, 1, 18, 24, 38–40, 42, 45, 54, 55, 61, 76, 77, 79, 80, 87–91, 93–95, 101, 106, 114, 120, 127, 129–132, 135.

occur, 76, 84, 94, 98, 100, 105, 107.
occurrence, 88, 98, 104.
offset, 46, 47.
opaque, 24.
opcode, 23, 33, 34, 45–47, 50, 58, 134.
OpenVera, 99, 100.
operand, 23, 45, 46, 54, 67, 71, 73, 93, 134, 149.
operate, 40–42, 112, 113, 140.
operation, 46, 76, 80.
operational, 2, 91.
operator, 16, 84, 94.
optimization, 76.
optimize, 57, 76, 94, 116, 148.
option, 55, 99, 119, 143, 170, 171, 191.
optional, 92, 99.
order, 2, 3, 5, 15, 18, 27, 28, 31, 35, 37, 61, 64, 78, 80, 95, 106, 112, 114, 116, 120, 124, 127, 128, 131, 134, 135, 137, 142, 148, 150.
organization, 26, 79, 99.
organize, xiii, 90, 111, 114, 148.
original, 17, 19, 100, 103, 136.
orthogonal, 20, 31, 35, 37, 75, 140.
outline, 15, 30, 76, 111.
output, 25–27, 37–39, 45, 55, 62–66, 68, 112, 114, 120–122, 130–132, 139, 141.
outside, 16, 113.
OVA, 99–101.
overflow, 21, 68, 100, 105, 141, 144.
overhead, 81.
overlap, 16, 80, 140, 147, 150.
overview, xiii, 78.
OVL, 99, 103, 141, 144.
oxymoron, 25.

P

package, 26, 151, 153.
packet, 23, 25–27, 34, 45, 106, 107, 130.
page, i, viii.
paragraph, 32, 36.
parallelize, 80.
parameter, 2, 31, 33, 37, 41, 45, 99, 112, 119, 130, 143, 156.
parity, 135.
parse, 36.
part, 29, 110.
partition, 20–22, 100, 109, 134, 135.
pass, 19, 27, 88, 102, 120, 126–128, 141.
passive, 63.
past
  bugs, 29.
  coverage, 130.
  simulation, 129.
  success, 110.
  visitors, xi.
patent, 23.
path, xi, 27, 52, 55, 80, 83, 94, 106, 107, 129, 130, 140, 141, 143, 144, 147.
PE, 49, 51.
Pedneau, Mike, xv.
Peled, Ofer, 76.
perform, 3, 19, 28, 29, 61, 90, 103.
performance, 23, 69, 76, 80, 110, 113, 128, 133, 148.
period, 27, 29.
periodical, 114, 130.
permutation, 28, 37, 39, 43, 45, 54, 57, 58, 84, 95, 107, 139–141, 144.
permute, 52.
perspective, xv, 24, 38, 39, 67, 89, 149.
PF, 58, 61, 68–70, 135.
phase, 44, 124.
pin, 27, 28, 49, 63, 64, 94.
pipe, 40, 114.
pipeline, 24, 34, 36, 37, 114.
place, xi, 4, 35, 38, 66, 104, 106, 111, 113, 120.
plan, xi, 19–24, 26, 29, 30, 101, 109–112, 114, 121, 123, 124, 129, 147, 148.
plane, 135.
planning, xi, 109, 111.
platform, 128.
point, xv, 4, 42, 51–53, 55, 57, 58, 78, 88, 89, 105, 114, 124, 127–129, 131, 134, 139, 140.
pointer, 46, 50.
populate, 71.
population, 37.
port, 69, 100, 143, 147, 178.
posedge, 101.
position, 44, 117, 133.

positive, 99.
possibility, 92, 105.
possible, 1, 19, 22, 26, 30, 41, 46, 48, 56, 83, 90, 92, 114, 127, 139, 148, 149.
postpone, 109.
practice, 3, 129.
pragma, 75, 91.
preserve, 15, 17, 18, 97, 126.
principle, x.
print, 168.
priority, 45, 91, 140.
private, 153.
probability, 92, 110, 114, 115, 129, 130.
problem, xiii, 5, 18, 20, 22, 30, 31, 88, 89, 109–111.
problematic, ix.
procedural, 22, 63, 68, 98, 101, 106, 107.
procedure, 76, 103, 148, 159.
proceed, 114, 141.
proceeding, 112, 120.
process, ix–xi, xiii–xv, 1, 15, 18, 19, 24, 25, 27, 28, 30, 37, 39, 40, 44, 46, 61, 75, 77, 78, 81, 89, 97, 103, 107, 109–111, 114, 123, 129–131, 134, 137, 139, 140, 149.
processing
    latency, 34.
    packet, 27, 106.
    text, xv.
    throughput, 34.
processor, 22, 26–28, 45, 49, 51, 57, 63, 64, 110, 140, 141.
product, 18, 42, 78, 89, 123.
production, 130, 131.
program, 22, 36, 46, 63, 75, 76, 79, 80, 88, 89, 91, 93, 103, 116, 117, 141, 143.
programming, xv, 65.
progress, ix, xi, xiii, 2, 19, 22, 30, 31, 37–39, 78, 109, 111, 127–129, 147, 149, 150.
    RTL coding, 123.
project, xi, 110, 112, 135.
projection, 134, 135.
proof, 1, 3, 77, 78, 104, 149.
propagate, 87, 88, 122.
properties
    interface, 105.
    protocol, 105.
    structural, 105.
property, 1–4, 77, 78, 97–101, 104, 105, 107, 114, 149.
protect, 46, 49, 55, 57, 67, 134, 140, 141, 153.
protocol, 23, 26, 64, 105, 106, 114.
    properties, 105.
prove, 1–4, 77, 78, 97, 102, 104, 149, 150.
proximity, 134.
prune, 23, 37, 131, 148.
PSL, 99–101, 103, 149.
purposeful, 50.

Q
quality, 128.
quantify, 33, 45, 77, 78, 110, 123, 131.
quantitative, 19, 77, 135.
question, ix, x, xiv, 19, 61, 65, 66, 75, 113, 141.
queue, 36.
quiescent, 90.

R
ranch, 29, 124.
random, ix, 22, 29, 90, 113, 115.
range, 37, 40, 46, 47, 114, 116, 118, 119, 124, 135, 136, 192.
ranging, ix, 40.
rare, 3, 4, 17, 91, 92.
rate, ix, xiv, 3, 90, 116, 122, 123, 127, 128.
ratio, 102, 127.
readable, 18, 34.
reality, 17, 18, 51.
realization, 18.
realize, 75.
reason, xiv, 26, 75, 80, 84, 91, 92, 110, 112, 128.
recognize, 36, 64, 75, 87.
record, 2, 33, 35, 41, 45, 46, 48, 54, 56, 57, 61–64, 72, 74–78, 80, 84, 86, 88–90, 93–95, 101–104, 109, 111, 113–115, 121, 123–126, 131, 137, 139–142, 146, 148, 149.
recursive, 24.
redundancy, 17, 18, 115.

redundant, xi, 115, 116, 129, 147.
reference, x, xv, 5, 22–26, 45, 47, 68–70, 112, 122, 124.
refine, 18, 66, 67, 124, 131.
refinement, 17, 67, 130, 132.
reflect, 36, 40, 50, 53–55, 59, 75, 76, 78, 139.
region, 16, 17, 34, 38, 52, 55, 111, 114, 130, 131, 133, 134, 137, 139, 140.
register, 1–3, 21, 33, 38, 40, 45–50, 54, 57, 58, 61, 64, 67–71, 73, 76, 79, 84, 85, 93, 94, 134, 141.
    coverage model, 71.
    operand pairs, 54.
        specification, 54.
        structure, 54.
regress, 28, 29.
regression, 19, 28–30, 88, 105–107, 114, 124.
regular, 55, 103.
reintroduce, 29.
reintroduction, 30.
relate, 37, 50, 56, 66, 78, 81, 83, 131, 140.
relationship, 36, 40, 41, 43, 44, 49–59, 67, 76, 78, 92, 93, 114, 121, 124, 131, 132.
release, 78, 90, 104, 161.
relevant, 130.
reliable, 110.
repeat, 164.
replicate, 79, 80.
report, xiv, 24, 35, 37, 38, 77–83, 85–89, 91–94, 98, 101, 104, 106, 113, 126, 131, 141, 147.
represent, 16, 17, 37, 52, 55, 57, 59, 78, 101, 131–134, 136, 137, 140.
representation, 1, 22, 32.
reproduce, 59, 70, 91.
request, 23, 28, 37, 98, 101.
require, xiii, xiv, 3, 17, 20, 22–25, 27, 30, 37, 41, 44, 54, 55, 61, 62, 64–67, 77, 80, 84, 90, 92, 94, 95, 107, 111, 112, 120, 122, 126, 129–131, 140, 141, 148–150.
requirement, 3, 4, 18–21, 23–25, 39, 40, 45, 50, 62, 90, 97, 98, 101, 109, 110, 112–114, 116, 120, 121, 125, 129, 131, 132, 137, 140, 147.
    source, 21.
requisite, 55, 127.
research, 76, 120, 131, 149.
reserve, 48, 68.
reset, 3, 4, 27, 28, 98, 101, 104, 146.
resistance, 55, 110.
resource, xi, 91.
respect, 115.
response, xiv, 2, 19–21, 24, 25, 27, 30, 37, 39, 64, 65, 68, 89, 109, 112, 120, 127, 137, 139.
responsibility, 22, 63, 109, 112.
responsible, 21, 24, 63–65, 88, 101, 104, 105, 120, 137.
restrict, 23, 27, 43, 88, 90, 116, 132, 149.
restriction, 28, 43, 73, 114, 132, 135, 136.
result, ix, 17, 21, 24, 25, 35, 37, 76, 78, 91, 94, 95, 97, 103, 106, 107, 110, 112, 120, 122, 126, 130, 131, 137, 139, 140, 148.
resultant, 107, 111.
RET, 141, 147.
retire, 50.
retrieve, 62.
return, 29, 37, 57, 67, 91, 92, 167.
reusability, 26.
reusable, 26, 30.
reuse, 26, 62, 110.
reveal, 18, 57, 90, 106, 125, 134, 149.
reverse, 4, 93, 164.
review, ix, 28, 78, 88, 106, 120, 129–131.
reviewer, xv.
rework, 110.
rigorous, 18, 19, 77, 88, 149.
risk, xiv, 90, 91.
risky, 112.
routine, 152, 155, 159.
row, 41, 42, 56, 58.
RSM, 49.
RTL, xiv, 1, 2, 16, 18, 22, 27, 31, 32, 34–37, 55, 79–81, 85, 88–92, 95, 97–99, 101–104, 106, 107, 109, 110, 121–129, 141, 147–149.
    change rate, 123.
    coding progress, 123.
RTPG, 22.
rule, 23, 40, 41, 105, 118.

run, 88, 107, 128.
runtime, 120.

S
safety, 97, 98.
sample, 11, 19, 41, 43, 45, 47–51, 56, 58, 61–67, 69, 70, 75, 90, 99, 100, 119, 141–144.
sampling
input attributes, 62.
internal attributes, 65.
output attributes, 64.
sampling event, 11.
sanity, 125.
Santini, Simone, 34.
SAT, 2.
satisfactory, 124.
satisfiability, 2.
satisfy, 104.
scarce, 91.
scenario, 38, 50, 88, 95, 104, 109, 112, 113, 130, 132, 139, 140, 148.
schedule, 110, 112.
Schwartzburd, V., 22.
scope, 20, 28, 30, 39, 100, 109, 111, 120.
scoreboard, 25, 26, 66.
search, 3, 4.
seed, 29, 90.
select, 33, 34, 37, 46, 48, 49, 55, 56, 66, 67, 78, 89, 91, 94, 98, 100, 114–116, 133, 178, 180.
selection, 29, 46, 178, 179.
semantic, 18, 27, 35, 36, 41, 44, 45, 54, 57, 67, 77, 94, 107, 135, 141, 149.
sensitize, 89.
sentence, 32, 36.
separation, 63.
sequence, 1, 4, 23, 27, 45, 81, 86, 100, 101, 104, 107, 117, 120, 130, 139, 153.
sequential, 2, 37, 54, 76, 82, 85–87, 93–95, 103, 148.
SERE, 103.
session, 45.
SF, 58, 61, 68–70.
share, xiii, 72, 121, 131, 137.
Shiell, Jon, xv.
shortcut, 120.
SI, 68.
silicon, xiii, 88.
similar, xi, xiii, 77, 135, 148, 149.
similarity, 2, 135.
simple, 3, 40, 43, 78, 100, 102, 119, 143, 178.
simplification, 51.
simplify, 55.
simply, 27, 49, 102.
simulate, 2–4, 27, 29, 30, 35, 90, 97, 102, 104.
simulation, xv, 3, 4, 18–20, 22, 24, 27–30, 80, 81, 88–90, 94, 97, 103, 112, 114, 115, 128, 129, 148–150.
simulator, 4, 90, 102, 103, 111, 128, 141, 144, 159.
skepticism, 110.
skew, 115.
slice, 68.
slope, 123.
SmartMoney, 132, 133.
SMI, 49, 63, 64.
Smith, Gary, 19.
snapshot, 90, 104.
soft, 115, 116, 118, 188.
software, 2, 19, 45, 84, 110, 148.
Solidify, 78.
solution, 2, 3, 18, 106, 111, 140, 149.
solve, xiii, 30, 111.
solver, ix, 2, 23, 130.
source, 23, 31–33, 35, 39, 40, 46, 62, 79, 136, 137, 141, 142, 147, 148, 150.
SP, 68.
space, xiii, 3, 16, 17, 19, 23, 27, 31, 34–39, 42, 51, 52, 72, 79, 95, 110–112, 114, 123, 124, 130, 135, 139, 140, 148.
span, 81.
specification, xiv, 1, 16–21, 23–26, 28, 31–40, 45, 51, 54, 55, 57, 59, 62, 63, 67, 95, 97–101, 107, 113, 114, 116, 120, 121, 123, 131, 132, 135–137, 139, 140, 148, 149.
specify, 4, 17, 18, 20–22, 29, 30, 35, 36, 39, 40, 42, 44, 46, 47, 50, 54–56, 62–64, 68, 73, 74, 76, 80, 93, 95, 97, 100, 101, 105,

107, 109, 112, 114–120, 123, 135, 141, 143.
Specman, 78, 103.
speed, 128, 148.
SS, 134.
stable, 22, 103, 106.
stack, 140.
stage, 18, 49, 89, 114, 127, 131.
standard, 99, 111.
state, xiv, 2–4, 29, 34, 36, 39, 40, 51, 54, 57, 68, 69, 75, 76, 79, 85–87, 91, 94, 95, 97, 100–102, 104–106, 112, 115, 125, 127–129, 172.
statement, 32, 80–83, 88–94, 97, 98, 100, 118, 127, 128, 151.
static, 1–4, 18, 75, 77, 78, 94, 97, 101, 103–105, 140, 149, 150.
statical, 23.
status, 46, 48, 76.
step, xi, xiv, 20, 26, 27, 44–46, 50, 61, 67, 89, 90, 103, 107, 111, 134, 149, 165.
stepwise, 17.
stimulate, 2, 110.
stimuli, 2, 20, 23, 27, 28, 30, 39, 89, 92, 114, 115, 120, 131, 139.
stimulus, ix, xiv, 3, 19–24, 26, 27, 30, 39, 62, 65, 68, 105, 107, 109, 112, 113, 115, 120, 127, 129, 130, 132, 137.
storage, 46, 80.
store, 25, 26, 141, 147.
story, 45.
strategy, x, xiv, 24, 25, 64, 109, 142.
strength, 55.
stress, 2, 114.
Strickland, Mark, xv.
struct, 68, 70, 71, 75, 117–119, 146, 152.
structural, xiv, 2, 79, 105, 111, 136, 137, 140.
    properties, 105.
structural coverage, 35, 38.
structure, 24, 25, 32, 35, 36, 41, 42, 45, 51–55, 63, 67, 68, 70, 75, 76, 78, 111, 130, 136, 141.
    coverage model
        register operand pairs, 54, 71.
subexpression, 32, 85, 90.
subgroup, 75.
subject, xiii, xiv, 5, 22, 24, 26, 28, 35, 102, 112–114, 140.
subregion, 52, 53.
subset, 16, 19, 23, 46, 58, 59, 77, 91, 114, 115, 139, 149.
subsystem, 149.
subtype, 65, 66, 69–71, 73, 75.
succeed, 67, 102, 145.
success, 110, 146.
successful, 17, 27, 64, 120.
Sugar, 100, 103.
superscalar, 88.
superset, 77.
supply, 17.
support, ix, 36, 63, 69, 89, 101, 110, 111, 139.
suppress, 124.
SureCov, 103.
SVA, 99.
symbolic, 22, 149.
symmetric, 55, 80.
symmetrical, 29, 30.
symmetry, 54.
sync, 172.
synchronous, 106.
synonymous, 121.
Synopsys, xi, 89.
syntactic, xiv, 36.
syntax, 5, 99.
synthesize, 104.
SystemC, 61.
SystemVerilog, 18, 35, 61, 84, 93, 94, 99, 101, 103, 111.

T
task, 67, 159.
taxonomy, x, 31, 37, 38, 140.
    coverage metric, 32.
technique, ix, 2, 3, 22, 75, 78, 79, 111, 122, 129, 131, 136, 148, 149.
temporal, xiv, 11, 23–25, 28, 35, 39, 67, 68, 75–78, 89, 100, 101, 104, 105, 114, 120, 121, 130, 137, 145, 146.
    coverage, 11.
test, ix, xiii, 12, 15, 19, 22, 27, 29, 30, 76,

88–90, 92, 93, 99, 109, 110, 120.
TF, 135.
The Visual Display of Quantitative Information, 134.
theorem, 1, 2, 149.
theory, 17, 18.
thread, 98, 141.
threshold, 80, 90.
time, xiii, xv, 2, 3, 22, 25–29, 37, 40–42, 45, 48–51, 53, 55, 56, 58, 61–63, 66, 67, 69, 70, 75–77, 79, 80, 83–86, 88, 90, 91, 94, 97, 98, 100–103, 105, 106, 110–112, 115, 116, 119, 120, 122–124, 128, 130, 131, 139–141, 160.
timing, 1, 24.
toggle, 85, 94, 95.
  coverage, 12.
toggle coverage, 85.
topological, 1.
trace, 89, 130, 148.
track, 24, 64, 111, 128.
trajectory, 149.
transaction, 36, 37, 64, 148.
transfer, 1, 38, 82, 92, 93.
transform, xi, 17, 25–27, 131, 135.
transformation, 18.
transition, 27, 85, 87, 94, 100, 170, 173.
translate, 18, 33, 146.
translation, 76.
transmit, 18.
transparent, 24.
trap, 135.
traversal, 94, 95, 125.
traverse, 77, 86, 94, 95, 104, 107, 124, 127, 128, 143, 148.
trigger, 84, 94, 99, 102, 103.
tristate, 85.
type, 34, 35, 51, 52, 68, 69, 71, 73, 75, 92, 100, 106, 107, 113, 115, 117, 118, 123, 134, 140, 143, 147, 150, 153.
typographical error, 1.

U
uint, 43, 115–117.
uncertainty, xiii.
uncomfortable, 110.
unconditional, 93, 106.
undefined, 59, 85, 155, 179.
understand, x, 15, 17, 78, 95, 111, 114, 134.
understanding, xv, 95, 97, 111, 133.
undetected, xiii, 29.
unevaluated, 101.
unexercised, ix, xi, 91.
unexpected, 106.
unfamiliar, xi, 5.
unforeseen, 4, 113.
uniform, 114.
unify, 34, 150.
unimplemented, 16, 17.
unintended, 16, 17, 132.
union, 16.
unique, 72, 90, 117, 129, 143, 149.
unit, 2, 63, 65, 66, 68, 75, 100, 106, 107, 114, 116, 121, 142, 143, 149, 158.
unobserved, 94.
unrelated, 65.
unseen, 113.
unsigned, 47.
unspecified, 16, 17, 24, 36, 64.
unusual, 90.
unwilling, 18.
update, 18, 48, 49.
Ur, Shmuel, xv, 131.
Uziel, Shlomi, xv.

V
valid, 23, 41, 43, 62, 63, 76, 114, 115, 131, 132.
validate, 30, 104.
validation, 1, 76.
validity, 3.
valuable, 78, 87.
value, xi, 28, 31, 34, 35, 37, 40–42, 45–53, 55–58, 61, 63, 64, 68–74, 76, 77, 84, 85, 88, 90–94, 99, 100, 105, 110, 113, 114, 116, 118–121, 124, 128, 129, 132, 134–136, 139, 142–144, 146.
variable, 61, 76, 93, 159.
variant, 78, 80, 110.
VCS, 89.
vector, 3, 22, 27, 104.
verification, ix–xi, xiii–xv, 1–3, 5, 12, 15,

17–24, 26–32, 37–39, 44, 45, 61, 62, 64–68, 75–79, 85, 88–92, 97–101, 103, 109–115, 120–131, 136, 137, 139–141, 147–150.
coverage-driven, 109.
functional, 15.
verification interface, 12.
verify, ix, xi, 1, 3, 4, 12, 18–21, 23, 24, 27, 29, 35, 76, 88, 97, 100, 111, 114, 128, 150.
Verilog, 18, 32, 35, 61, 81, 84, 93, 94, 99, 100, 143, 144, 159, 160.
Verisity Design, 23, 26, 78, 103, 110.
Verplex, 99.
VHDL, 18, 32, 35, 61, 80, 82, 84, 93, 94, 159, 160.
view, 94, 131, 147.
violate, 1, 3, 4, 104.
violation, 101, 104, 105.
virtual, 51, 55, 67, 139, 140.
virtual-8086 mode, 56, 57.
visibility, 78, 80, 81, 127.
visit, xi, 45, 85, 86, 94, 104, 112, 114, 125, 127–129.
visualization, 132–134.
visualize, 98, 130.
VM, 51, 55–57, 67.
volatile, 64, 89.
volatility, 64, 123.
volume, 45.

W
weight, 12, 114, 115.
weighted, 129.
weighted average, 12.
white, 24, 122, 133, 134, 150.
whiteboard, 15.
wire, 84, 85.
Writing Testbenches, xiii.

Z
ZF, 58, 61, 68–70.
Ziv, Avi, xv, 76, 120, 131.